a

Rocks of this age and older Precambrian rocks probably underlie the Triassic Knowle Basin.

Following uplift and erosion of the district in Ordovician time, the earliest Silurian sandstones and mudstones, represented by the **Rubery Formation**, were deposited unconformably on the **Lickey Quartzite**. During Wenlock and Ludlow times, carbonates, including patch reefs, accumulated in shallow warm seas. Siliciclastic rocks with fish bone beds were deposited in late Silurian and early Devonian times, in shallow marine and eventually in brackish water, and finally, with a marine regression, in continental conditions. Caledonian (Acadian) earth movements affected the area prior to a long period of erosion.

During Carboniferous times the district lay at the southern margin of the Pennine Basin, to the north of the Wales–Brabant High. Dinantian and Namurian strata are absent. Westphalian **Coal Measures** rest unconformably on rocks of Cambrian to Devonian age. Sandstone, mudstone and coal were deposited in deltaic and lacustrine environments. The Coal Measures are relatively thin in both the South Staffordshire and Warwickshire coalfields, but includes the Thick Coal of great economic importance; ironstone and fire clay were also exploited. Increasingly well-drained conditions on the Carboniferous floodplain and increased sediment input in Bolsovian times, resulted in an upward diachronous change to red-bed deposition of the **Etruria Formation**, a prime source of brick clay. Dolerite bodies, including the **Rowley Regis**

lopolith, were intruded at this time. Variscan earth movements resulted in folding and reverse faulting prior to deposition of the **Halesowen Formation** in Westphalian D times. The Halesowen Formation includes thin coals representing a return to humid deltaic conditions. Above it, the Carboniferous rocks are red, poorly fossiliferous, fluvial and lacustrine deposits; they include conglomerates with clasts of Carboniferous limestone derived from newly uplifted source areas.

Uplift during the early Permian resulted in unroofing of Lower Palaeozoic and Precambrian rocks in the source lands and deposition of **Clent Formation** breccias in proximal alluvial fans; these pass basinward into sandstones and mudstones. The **Hopwas Breccia** contains clasts derived from a variety of hinterland sources, and was deposited in similar environments, probably in early Triassic times. During Permian times a change in the style of deformation from the compression/transpression of the Variscan orogeny gave way to a phase of regional east-west extension. The Knowle Basin is one of a series of graben and half graben that developed during this period.

A thick succession of Triassic rocks, the **Sherwood Sandstone Group** was deposited in fluvial environments, infilling the rapidly subsiding basin. Thinner successions were deposited on the adjacent highs. The overlying **Mercia Mudstone Group** was deposited in fluvial and lacustrine (playa) environments; gypsum and desiccation features testify to a semi-arid climate.

Towards the end of Triassic times an extensive transgression occurred; it is represented by fossiliferous mudstones and limestones of the **Penarth Group**, which together with the overlying marine Jurassic rocks crop out in a small fault-bounded outlier.

Quaternary deposits (till, sand, gravel and clay), mostly products of the Anglian glaciation, and periglacial deposits cover a large part of the district; organic-rich Hoxnian interglacial deposits have been proved in boreholes. Palaeovalleys (buried channels) were cut into the bedrock and infilled with diverse glacigenic and interglacial deposits. Younger drift deposits, including mass movement deposits (head), are widespread. The Quaternary deposits reflect the fluctuating climate during the past two million years.

A synthesis of the tectonic history and deep structural data derived from geophysical data is presented, and the reader is provided with a section listing information sources.

Cover photograph   View of central Birmingham, underlain by the Sherwood Sandstone Group, looking north-east along the Grand Union canal (MN 28018) (Photographer: T Cullen).

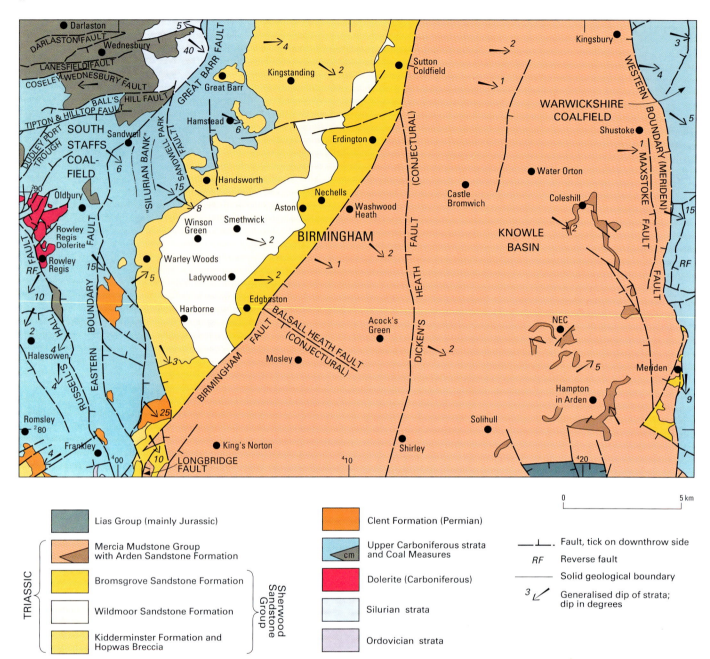

Figure 1 Sketch map of the solid geology of the district.

BRITISH GEOLOGICAL SURVEY

J H POWELL
B W GLOVER
C N WATERS

Geology of the Birmingham area

CONTRIBUTORS

Biostratigraphy and sedimentology
C R Hallsworth
A W A Rushton
G Warrington

Engineering geology
P R N Hobbs

Geophysics and seismic interpretation
C P Royles
N J P Smith

Stratigraphy
D A Piper

Hydrogeology
J Davies

Memoir for 1:50 000 Geological Sheet 168
(England and Wales)

London: The Stationery Office   2000

The grid used on the figures is the National Grid taken from the Ordnance Survey map. Figure 2 is based on material from Ordnance Survey 1:50 000 scale map, number 139.
© Crown copyright reserved.
Ordnance Survey Licence No. GD272191/2000

ISBN 0 11 884545 4

Bibliographical reference

POWELL, J H, GLOVER, B W, and WATERS, C N. 2000. Geology of the Birmingham area. *Memoir of the British Geological Survey*, Sheet 168 (England and Wales).

Authors

J H Powell, BSc, PhD, CGeol
C N Waters, BA, PhD, CGeol
British Geological Survey, Keyworth

B W Glover, BSc, PhD
formerly British Geological Survey, Keyworth

Contributors

C R Hallsworth, BSc
P R N Hobbs, BSc, MSc, DIC
D A Piper, MA, PhD, CGeol
C P Royles, BSc
N J P Smith, BSc, MSc, CGeol
G Warrington, DSc, CGeol
British Geological Survey, Keyworth

A W A Rushton, BA, PhD
formerly British Geological Survey, Keyworth

J Davies, BSc, MSc, MGA
British Geological Survey, Wallingford

Printed in the UK for The Stationery Office
TJ001108 0 C6 3/2000

Acknowledgements

In this memoir, Chapter Two was written by Dr J H Powell, J Davies (hydrogeology) and P R N Hobbs (engineering geology); Dr A W A Rushton contributed Chapter Three; Chapter Four was written jointly by Dr C N Waters with assistance from Dr J H Powell and Dr A W A Rushton, Chapter Five was written jointly by Dr B W Glover and Dr J H Powell; Chapters Six and Seven by Dr J H Powell, with assistance from Dr G Warrington and Dr D P Piper; Chapter Eight by Dr J H Powell, with assistance from Dr C N Waters and Dr B W Glover. Chapter Nine was written by Dr C N Waters with assistance from N J P Smith and Dr J H Powell; C P Royles contributed the geophysical summary. Other parts of the memoir were written by Dr J H Powell who also compiled the text. BGS Technical Reports by the authors, and those by C R Hallsworth, Dr R J O Hamblin, Dr N S Jones, Dr G K Lott, Dr R A Old, Dr D P Piper, and M G Sumbler (Chapter Ten) have been freely used in this account, and except in a few special instances, are not specifically referred to in the text. Dr S A Johnson made available unpublished data on palaeo-magnetic analysis of Upper Carboniferous and Triassic rocks in the area. The memoir was edited by Dr A A Jackson and T J Charsley. The geological survey of the western (Black Country) and south-east (Coventry) parts of the district was in part supported by the Department of the Environment.

Thanks are due to the various quarry and gravel pit operators in the district who allowed access to their excavations, to British Coal and the Coal Authority for providing details of boreholes, and to the Severn-Trent Region of the Environment Agency (formerly National Rivers Authority) for providing licensed abstraction data. We thank Dr P Smith for facilitating access to the palaeontological collections at the University of Birmingham.

Notes

Throughout the memoir, the word 'district' refers to the area covered by the 1:50 000 Series Birmingham Sheet 168.

National Grid references are given in square brackets; those beginning with the figure 9 lie in the 100 km square SO, and those beginning with the figures 0, 1 or 2 lie in the 100 km grid square SP. Alphabetical prefixes are given for localities outside the Birmingham district.

Numbers preceded by the letter E refer to thin sections in the National Sliced Rock Collection at BGS, Keyworth.

A list of boreholes quoted in the text is shown in Chapter Ten.

CONTENTS

FIGURES

PLATES

TABLES

PREFACE

The study of urban geology is becoming an increasingly important topic as the proportion of the world's population that lives in cities continues to increase. The BGS has been involved in the study of the major conurbations of Britain for many years and this study of the geology of the second largest city is of particular importance.

Geology has contributed much to the prosperity of the district and has had a profound influence on its development. Coal, ironstone and fireclay have been mined in the region for a considerable period and, along with the quarrying of clay for brick manufacture, dolerite for aggregate, sandstone, sand and gravel for aggregate and foundry sand, have contributed greatly to the wealth of the region. Groundwater, extracted principally from the Triassic sandstone, provided a major resource for industrial supply, but decline in abstraction over the past few decades has resulted in rising water tables in some areas which may have important consequences.

The Black Country to the west of the city earned its name in the 18th and 19th centuries from the pollution produced by the multitude of industries (collieries, blast furnaces and foundries) located on and near the Staffordshire Coalfield. Two centuries of mineral extraction and industrial development have left a legacy of land-use problems which can only be assessed by reference to geological maps and supporting databases. The geological map and memoir together provide the basic information required by the extractive industries and construction industries for sustainable exploitation of mineral resources. Key issues which should be given a high priority in land-use planning and development include the location of past mineral workings and undermined areas, sterilisation of resources under new development, water resources including aquifer vulnerability and pollution, foundation conditions, landslides, the extent of artificial deposits, landfill, the generation of explosive and toxic gases, and conservation.

David A Falvey, PhD
Director

British Geological Survey
Kingsley Dunham Centre
Keyworth
Nottingham
NG12 5GG

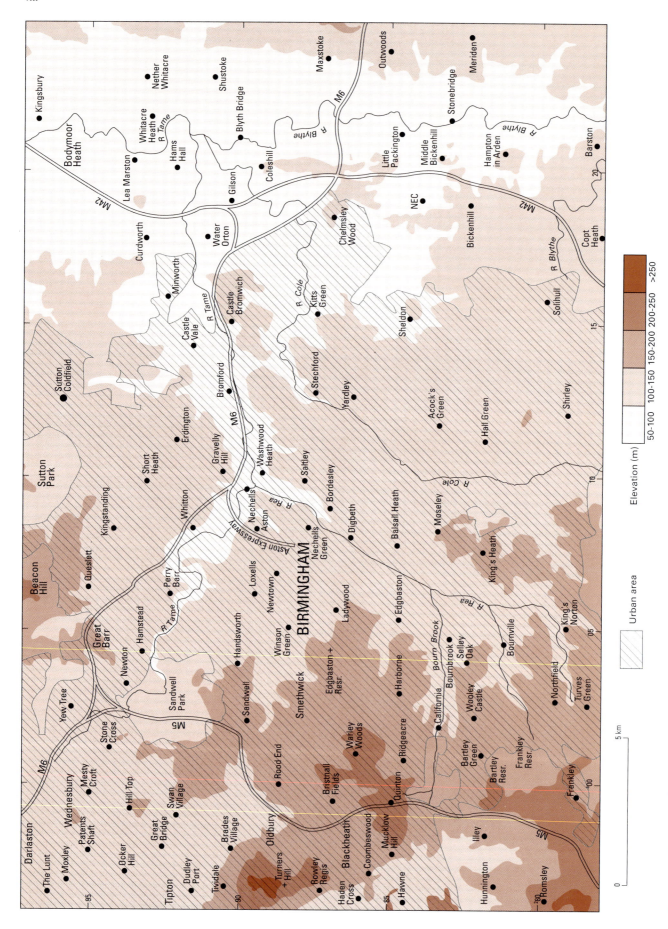

Figure 2   Topography of the district.

ONE

Introduction

The district around Birmingham (Sheet 168), described in this memoir, forms part of the West Midlands conurbation; it includes the City of Birmingham, part of the counties of Warwickshire, Worcester and Staffordshire, and parts of the metropolitan boroughs of Sandwell, Walsall, Wolverhampton, Walsall and Dudley. The district is mostly urban but includes some agricultural land to the north, east and south; it includes the eastern part of the South Staffordshire Coalfield and the western part of the Warwickshire Coalfield (Figure 1). Coal, ironstone and fireclay mining and brick manufacture, formerly important industries in parts of the district, have ceased, and mineral extraction is presently restricted to the quarrying of aggregate, including dolerite and Quaternary sand and gravel.

PHYSIOGRAPHY

The topography of the district (Figure 2) generally reflects the underlying geology. The central part, which largely comprises the Triassic Knowle Basin, is bounded to the west by the Eastern Boundary (Great Barr) Fault which defines the eastern margin of the South Staffordshire Coalfield. The eastern margin of the Knowle Basin is defined by the Western Boundary Fault of the Warwickshire Coalfield. The Knowle Basin is underlain by Triassic rocks concealed for the most part by a blanket of drift deposits, and lies at an elevation of between 120 and 150 m above OD. To the north-west of the Birmingham Fault the higher ground in central Birmingham is underlain by harder Triassic sandstones, mostly obscured by drift deposits.

The highest ground, in the south-west, forms part of a prominent north-east-trending ridge that extends from Sedgley, beyond the district in the north-west, through the Silurian limestone ridges, to the Rowley Regis Hills (267 m), Quinton (224 m), Perry Hill (219 m) and Warley (219 m). South of Halesowen, the ground rises again to form a ridge between Frankley Beeches (250 m) and Romsley Hill, in the south-west, which marks the highest elevation at 283 m above OD. The ridge, which represents part of a major watershed in the district and of central England, is broadly defined by the trace of the Russell's Hall Fault, and is largely due to the presence of hard, resistant strata such as intrusive dolerite, to the north-east of the fault, and the Permian Clent Formation, to the south-west. The relative hardness of the Triassic strata, the Hopwas Breccia and Kidderminster Formation, gives rise to a series of ridges at Warley, and farther north between Perry Barr and Barr Beacon where the ground rises to about 220 m. In general, however, the Triassic Sherwood Sandstone is typified by gently undu-

lating topography, such as the well-drained area from Harborne through Edgbaston to Handsworth. Gently undulating topography is also characteristic of the drift-covered areas, typified by the low-lying region around Wednesbury, and to the east of the coalfield, between Smethwick and Handsworth.

To the south-west of the ridge that forms the major watershed, streams drain to the west, including the River Stour which rises in the Romsley area, and flows via the River Severn to the Bristol Channel; to the east of the ridge, the Tame and its tributaries the Rea, Bourn Brook, Hockley Brook and Plant's Brook, generally flow eastward. Downstream of the confluence with the northward flowing Cole and Blythe, and the westward flowing Bourne, near Hams Hall, the Tame turns sharply northward, eventually flowing via the River Trent to the North Sea. Parts of the river courses have been culverted, and their valleys largely exploited by the extensive network of canals (including the Grand Union and Birmingham–Fazeley canals) and railways, which formed the principal communication routes until the rise of the urban motorways.

GEOLOGICAL HISTORY

The oldest known rocks in the district are Lower Palaeozoic in age. They were deposited on the Midland Microcraton to the east of the rapidly subsiding Welsh Basin. A small outcrop of the Lickey Quartzite, located in the south-west, is probably Ordovician in age; it was deposited in shallow marine conditions, subsequent to the volcanism which produced the underlying Barnt Green Volcanic Formation (Tremadoc), proved in the sheet to the south. In the east of the district, rocks of Merioneth to Tremadoc age (Table 1) consisting of dark grey and pale grey mudstones of the Monks Park Shale and Merevale Shale formations, respectively, have been proved, in boreholes, below the Coal Measures. These shaly formations (Stockingford Shale Group) contain a sparse graptolite, trilobite and brachiopod fauna, and were probably deposited in an epeiric shelf sea; the base of the Merevale Shales marking more oxygenated bottom conditions.

Following earth movements in the late Ordovician, the earliest Silurian strata represented by the late Llandovery Rubery Formation were deposited unconformably on the Lickey Quartzite. The Rubery Sandstone and overlying Rubery Shale record a marine transgression with an upward diminution of coarse-grained clastics as the Silurian sea spread eastward across the Midland Platform. The formation was deposited in a warm, shallow shelf-sea rich in shelly fauna, such as brachiopods, trilobites and corals. These beds were succeeded in Wenlock times, by shallow-water mud and silt, and this type of sedimentation continued throughout most of the Silurian. However, at times the supply of siliciclastic sediment diminished and, during slight falls in relative sea level, carbonates such as the Barr Limestone and Much Wenlock Limestone were deposited in shoaling conditions on the platform. The limestones are composed of abundant and varied shelly fossils including trilobites, crinoids, brachiopods and bivalves. Small patch-reefs, rich in corals, stromatoporoids and algae, are present in the Much Wenlock Limestone, and contemporaneous bentonite beds (water-laid tuffs) indicate distant volcanism during Wenlock and early Ludlow times. In this district, the Much Wenlock Limestone and overlying Silurian and Devonian strata are known only from boreholes.

Regression of the sea in late Silurian to early Devonian times, associated with the Acadian (late Caledonian) orogeny, resulted in a change of sedimentation as streams fed sand and silt into the basin during a gradual transition from marine, through brackish, to a continental environment. The continental deposits include fluvial sediments, deposited on a coastal plain, with some winnowed, condensed beds rich in the remains of fresh-water fish and *Lingula*.

During the Caledonian earth movements, the district was subjected to tilting, faulting, gentle folding, uplift and erosion. Subsequent to these earth movements, sedimentation resumed in late Carboniferous times when the Coal Measures of Langsettian, Duckmantian and early Bolsovian age (formerly Westphalian A to C) were laid down, predominantly in delta-plain, lacustrine and swamp-like environments over the eroded, irregular floor of Silurian and Devonian rocks; steep-sided highs occur locally on this floor. The district lay in tropical latitudes, and close to the southern margin of the Pennine Basin. The latter approximates to the area around the Clent and Lickey Hills which formed part of the Wales–Brabant Massif (land barrier). Sediment, derived mostly from the north (Scandinavia), was deposited by southward flowing rivers and deltas; thickly vegetated peaty swamps developed on the floodplain. Occasionally the Pennine Basin was invaded by the sea during relatively high sea-level stands, resulting in the deposition of marine silt and mud; the resultant shales (marine bands) can be correlated over wide areas.

Fluctuations in the rates of subsidence and sedimentation, and fluctuations in worldwide (eustatic) sea levels are reflected in the distinct cyclicity of the Coal Measures. A typical, complete cycle begins with marine or brackish shales (such as the named marine bands) which pass up into lacustrine and deltaic deposits consisting of mudstones, siltstones and sandstones. Colonisation of the delta-top by plants during periods of relatively reduced subsidence resulted in the formation of thick peat, underlain by leached soil (seatearth fireclay). Thick seams such as the Staffordshire Thick Coal (up to 12 m thick) represent such sites; this seam splits into a number of thinner seams to the north reflecting increased rates of subsidence towards the basin centre.

During late Bolsovian times, a gradual transition to well-drained, fluvial conditions is manifested in the red-bed Etruria Formation. Mud, generally devoid of carbonaceous material, was deposited in oxidising conditions upon an alluvial floodplain which was periodically cut by fluvial channels in which coarse-grained, pebbly sand was deposited. Dolerite, in the form of dykes, lopoliths and sills (Rowley Regis), was intruded into the Carboniferous strata in Bolsovian times when these sediments were buried at shallow depths. Following a phase of uplift and erosion in late Bolsovian times, the grey beds of the Halesowen Formation mark a brief return to sedimentation in alluvial and lacustrine environments, similar to the Coal Measures.

In late Carboniferous (Westphalian D to Stephanian) and early Permian times, deformation, uplift and faulting of the highlands to the south of the Pennine Basin, and possible inversion of the Knowle Basin, resulted in rejuvenation of rivers which deposited pebbly sand, silt and mud on broad alluvial plains in well-drained, oxidising conditions reflected in the red colour of the sediments (Salop Formation and Meriden Formation). In early Permian times, subsequent to the deposition of the Salop Formation, a period of erosion occurred which was followed by deposition of coarse-grained breccia, the Clent Formation. The breccia was deposited, in the south of the district, as alluvial fans at the foot of an emergent mountain front located to the south of the area. This lithology passes gradationally northwards into fine-grained sandstone, with pebbly lenses, and mudstone which was deposited in distal, more rapidly subsiding basinal areas to the north and north-east (Warwickshire). Later in Permian times, downfaulting of the area to the west of the district led to the development of the northern Worcester Basin in which aeolian sediments were deposited, but there is no direct evidence of these sediments within the district, although their presence has been postulated within the Knowle Basin.

Sedimentation in Triassic times took place in a semi-arid climate, and was controlled by east–west extensional tectonics. The South Staffordshire and Warwickshire coalfields were relative highs between which the subsiding Knowle Basin developed as an area of active sedimentation, with evidence of synsedimentary faults. The Hopwas Breccia is probably Triassic in age, and locally rests unconformably on late Carboniferous rocks; it includes pebbles of Lower and Upper Palaeozoic rocks and was probably deposited as alluvial fans infilling depressions in the local palaeotopography. The overlying Triassic Sherwood Sandstone Group is characterised by pebble-grade conglomerate, and pebbly sandstone at the base (Kidderminster Formation); far-travelled, well-rounded, exotic pebbles and boulders were deposited by northward flowing rivers. This passes gradationally upward to fine-grained sandstone and, less commonly, mudstone (Wildmoor Sandstone Formation), which were deposited in fluvial and lacustrine environments. The overlying Bromsgrove Sandstone is pebbly, in part, and was deposited in similar depositional environments; it locally oversteps the Wildmoor Sandstone to rest on the Clent Formation in the south-west of the district.

The overlying Mercia Mudstone Group consists of blocky red mudstone with thin, dolomite-cemented siltstone and sandstone beds, locally known as skerries; desiccation cracks and beds or nodules of gypsum indicate deposition in fluctuating fluvial and playa-lake environments in a semi-arid climate. The Arden Sandstone is a thicker sandstone-dominated unit in the upper part of the group and is characterised by grey-green colours; it contains a few fossils indicating a Carnian age, and was deposited in fluvial to possibly estuarine or marine conditions. The grey-green, dolomitic mudstones of the Blue Anchor Formation, which has a restricted marine fauna, reflects a change to a brackish, anoxic environment marking the onset of the Rhaetian transgression.

The Penarth Group records continuation of the widespread marine transgression in late Triassic (Rhaetian) time which resulted in deposition of black, carbonaceous mudstone and thin beds of bioturbated sandstone, with an abundant, low-diversity marine fauna, principally of bivalves (Westbury Formation). It is succeeded by the pale grey-green, calcareous mudstone (Cotham Member) which lacks a rich macrofauna, but contains abundant marine microfossils. The Penarth Group and overlying Lias Group have a small outcrop in the south-east of the district. Only the lowest part of the Lias Group, represented by the Blue Lias Formation (Hettangian Stage),

crops out in the district. It consists of grey calcareous mudstone with thin beds of shelly limestone, and was deposited in shallow marine conditions, which were established following a transgression of the Tethys Ocean over much of Britain during late Triassic to early Jurassic times.

A major phase of extensional deformation, in post-Jurassic times, resulted in the downfaulting of the Mesozoic strata along the principal faults, particularly those bounding the Knowle Basin. Bedrock younger than the early Jurassic were removed due to uplift and erosion during the Cainozoic.

The Quaternary drift deposits of the district include till, sand and gravel, silt and clay deposited during the Anglian and possibly the Wolstonian and Devensian glaciations, and Hoxnian interglacial, in environments ranging from glacial, periglacial, and fluvial to lacustrine, during climatic periods which fluctuated from glacial (very cold) to interglacial (temperate, warm). Locally, these deposits were laid down in steep-sided palaeovalleys (buried channels) cut into the bedrock. Head deposits, the result of gelifluction and solifluction processes, are present locally. During the Holocene, erosion modified much of the glacial landforms, and glacial deposits have been incorporated in alluvial deposits.

TWO

Applied geology

In this chapter the geological factors, relevant to land-use planning and development within the district are reviewed. The key issues are identified; some are considered in detail, but in all cases the reader is directed to sources where more information can be obtained.

KEY ISSUES

Geological factors have played an important part in shaping the industrial development of Birmingham and its surrounding areas (Powell et al., 1992). Nearly two centuries of mining, quarrying and industry have left a legacy of undermined ground, industrial dereliction, contaminated ground and pollution that has to be taken into account in land-use planning, and at all stages in the development and redevelopment of sites. The key issues given below are those which should be given a high priority in planning considerations (DETR, 1998). Table 2 lists some additonal published documents which provide advice and guidance on these planning and development issues.

- *mineral resources*: past exploitation, remaining resources, current activity, sterilisation by new development
- *water resources*: surface water, groundwater
- *fluctuating groundwater levels*: effects of reduced abstraction, predicted changes

- *aquifer vulnerability and pollution*: water chemistry, solvent contamination of groundwater, aquifer protection, leachate contamination from landfill sites
- *foundation conditions*: weathering zonation, rock strength, gypsum dissolution, running sands, ground stability, perched water tables, compressible superficial (drift) deposits
- *landslip*: past and potential mass movements
- *made ground*: extent, compositional variation, industrial contamination, spontaneous combustion, perched water tables
- *landfill*: gas generation, toxic residues, leachate mobility
- *undermining*: subsidence, shaft collapse, groundwater pollution, toxic gas generation
- *conservation*: protection of sites of scientific interest

MINERAL RESOURCES

Minerals have been exploited in the district for many purposes. They include: sand and gravel and crushed rock for construction; clay for the manufacture of bricks, tiles and pipes; coal for energy production; ironstone for iron and steel production, and limestone for flux and agricultural purposes. Options for quarry locations are

Table 2 Selected HMSO documents giving planning advice and guidance.

CIRCULARS (DoE, DETR)	
Pending	Development of contaminated land
27/87	Nature conservation
7/94	Environmental assessment
17/89	Landfill sites: development control
PLANNING POLICY GUIDANCE	
1.	General policy and principles (1997)
2.	Green belts (1995)
3.	Land for housing (1992)
4.	Industrial and commercial development (1992)
12.	Development plans and regional planning guidance (1992)
14.	Development on unstable ground (1990, 1996)
MINERALS PLANNING GUIDANCE	
1.	General considerations and the development plan system (1996)
2.	Applications, permissions and conditions (1988)
3.	Opencast coal mining (1994)
6.	Guidelines for aggregates provision in England and Wales (1994)
7.	The reclamation of mineral workings (1996)
12.	Treatment of disused mine openings (1994)
DERELICT LAND GRANT ADVICE	
1.	Derelict land grant policy (1991)

limited due to the nature and distribution of the strata, and quarrying may have adverse environmental effects such as noise and dust. Many of the pits and quarries have been backfilled, others partially filled and degraded, some are flooded while others remain open. Steep slopes associated with the excavation may create a risk of slope failure; problems associated with landfill are considered, below.

Information on the mineral resources of the district, interpreted from geological maps and records held by the BGS and other information sources (Table 2), should be consulted in order to make informed planning decisions on the environmental effects of mineral extraction, subsequent reclamation of the site, and on whether development might sterilise a potential resource.

Details of mineral resources in the Black Country area of the district (west of Easting [405]) can be found in Powell et al. (1992, and accompanying maps). Although a wide range of minerals has been exploited in the Birmingham area (reviewed below), the only recent (1995) workings are for dolerite, and sand and gravel, for use as aggregate.

Sand and gravel

Sand and gravel for use as aggregate, building sand and moulding sand was formerly worked in the district from both the superficial drift deposits and the Sherwood Sandstone bedrock, although recent workings are restricted to alluvium and river terrace deposits, and glaciofluvial sand and gravel, in the east of the district (see below).

DRIFT RESOURCES

The principal sand and gravel resources which have been exploited over the last decade are the **alluvium** and **First** and **Second river terraces** of the River Tame and its tributaries, the Rivers Cole and Blythe. Extensive shallow workings in these deposits are present in the Coleshill [205 905], Whitacre Heath [215 915] and Bodymoor Heath [205 950] areas (Figure 2). Workings generally extended downwards to the underlying Mercia Mudstone bedrock. Most of the former workings in this area have been backfilled, locally with pulverised fuel ash sourced from the former Hams Hall Power Station, but others have been allowed to flood so as to provide recreational lakes. First river terrace deposits of the River Blythe were formerly worked in the south-east of the district, near Ryton End (215 795). Details of the resources, including gradings, adjacent to the River Blythe, in the south-east of the district are given in Cannell (1982).

The sand and gravel of the above named river terraces and alluvial tracts generally consists of orange, yellow and red-brown clayey sand with gravel lenses, and coarse to pebble grade gravel. The sand is fine to coarse grained and consists mainly of subangular to well-rounded quartzite and subordinate quartz clasts (typical 'Bunter' suite). The gravel mainly consists of well-rounded pebbles and cobbles of quartzite (73–79%), quartz (19–25%) and sparse highly weathered intrusive rocks and quartz porphyry (Cannell, 1982). In the south-east of the district, mean gradings of the alluvium and river terrace deposits, respectively, are fines 5% and 9%, sand 27% and 58%, and gravel 68% and 33%. Locally, the river terrace deposits are highly cryoturbated, thereby incorporating clasts of mudstone bedrock, which reduces the quality of the resource. The maximum thickness of the units in this area is alluvium — about 3 to 4.5 m, first river terrace —3.5 m, and second river terrace — about 4 m. The most recent workings along the Tame valley (Harris et al., 1994) include those at Middleton Hall [195 975] in river terrace deposits, and at Coleshill [205 905], in both alluvium and river terrace deposits.

Glaciofluvial sand and gravel was extracted in a number of pits in the area, and the deposit includes areas of potentially workable sand and gravel (Cannell, 1982; Cannell and Crofts, 1984). In the south-east of the district the mean grading of the deposit is fines 12%, sand 63% and gravel 25%, but there is considerable vertical and lateral variability from pebble-free sand to gravel. The sand fraction is predominantly composed of fine- to medium-grade, angular to subrounded quartzite and quartz; the gravel fraction comprises rounded to well-rounded quartzite with subordinate quartz and sandstone and sparse mudstone and igneous rock clasts (Cannell, 1982). Recent (1995) extraction is centred on quarries [234 813 and 225 805] near Cornets End, Meriden. The deposit was formerly worked in the Arden area at two sites [223 824 and 230 827] near The Somers. Farther north, there were extensive workings at Middle Bickenhill [202 841], Little Packington [210 849], Great Packington [228 849], and between Gilson and Coleshill [192 904 to 196 900]. In the Shirley area the deposit was formerly extracted from scattered, small quarries at [117 800], [100 791] and [097 780].

In the west of the district glaciofluvial sand and gravel was formerly worked in scattered, small pits, such as those at Smethwick [SP 013 892], Stone Cross [SP 009 945] and Bustleholm [SP 015 945]. Other pits in the central part of the district include those at Washwood Heath [100 884], and Goosemoor Lane [108 934].

Glaciolacustrine sand and loamy sand, together with associated clay and silt, was worked in the Moxley area [971 952; 971 955; 966 965] (Powell et al., 1992) where the fine grain-size and low clay content made it ideal for use as moulding sand in local foundries; most of the pits in this area have either been wholly or partially backfilled, others are being utilised for recreational purposes. In the east of the district, glaciolacustrine sand was extracted at Dunton [190 932] (active in 1994) and Gilson [190 907].

BEDROCK RESOURCES

The soft, poorly cemented nature of the sandstones in the lower part of the **Sherwood Sandstone Group** led to their former exploitation for moulding sand, building sand and aggregate. These deposits are sometimes referred to as 'sand-rock'.

The lower pebble-rich part of the **Kidderminster Formation** was formerly exploited as a source of sand and gravel for aggregate. The most extensive workings were in the Queslett area [063 944] but these are now largely backfilled; additional, small pits include those at Pinfold

Lane Quarry [0592 9672], near Barr Beacon (where the underlying **Hopwas Breccia** was also worked), Warley [008 859], [001 869] and [014 866], Hamstead [046 949], and Harborne [015 842], [018 835].

The **Wildmoor Sandstone** was extensively worked for moulding and building sand in the Hockley area [0598 8815 to 0600 8806] and [0531 8854], where it was known as the 'Hockley moulding sand'; it was also worked in scattered, small pits across the outcrop, such as those near Smethwick [021 874] and [037 891], and at Selly Oak [042 829]. This formation was well suited to foundry purposes because of its uniform, fine grain-size and its slight clay content which rendered it readily mouldable.

The **Bromsgrove Sandstone** was locally worked, together with overlying Glaciofluvial Deposits near Perry Barr [098 900].

Dolerite (aggregate)

The Rowley Regis dolerite intrusion (see Chapter Five) is the chief source of road aggregate and crushed rock in the district, where it was locally known as 'Rowley Rag'. The dolerite intrusion is characterised by three distinctive joint patterns and by small faults that partly controlled quarrying operations. Dolerite was formerly used to make paving blocks, and was also admixed with concrete for the manufacture of concrete slabs, but most of the past and all of the present output is used in the production of crushed aggregate and coated stone.

Edwin Richards/Hailstone Quarry [969 883], the only recent working quarry, is about 100 m deep (Plate 1) and is utilised as a landfill site. Most of the former quarries around the Rowley Regis intrusion, such as Rough Hill [962 889], Yewtree Lane [967 869] and Bury Hill Park [977 891] have been partly or wholly backfilled. Further development of this resource is limited by urban sterilisation. Landfill gas (methane) generated at a nearby backfilled pit has been used as a fuel in the production of asphalt-coated roadstone.

Brick clay

The prime source of Brick Clay in the West Midlands is the **Etruria Formation** (locally termed the Etruria Marl or Old Hill Marl); it crops out in the west of the district and is still worked outside of the district in the 'Black Country' (Powell, et al., 1992), and to the north-east, near Kingsbury [219 993], in the adjacent Lichfield district. Brick manufacture was formerly carried out in the district exploiting the Etruria Formation in small pits scattered across the outcrop (Powell et al., 1992); these include workings at Old Hill [958 861], Round's Green [985 892], Mucklow Hill [980 928], Great Bridge [980 928] and Lyndon [005 925]. Resources have largely been sterilised by urbanisation.

The Etruria Formation is particularly suited to brick manufacture because the clays, mudstones and siltstones have a high iron oxide ($Fe^{3}+$) content which gives a characteristic red coloration (Holdridge, 1959). Lithological heterogeneity and colour variation in the brick clay

Plate 1
Carboniferous dolerite workings at Hailstone Quarry [969 883], Rowley Regis (1990). Note the prominent vertical joints (GS 710).

facilitates the blending of raw materials, which together with firing at variable temperatures, produces high quality bricks of various colours and hues. The high iron oxide content favours the production of the valuable engineering (blue) brick, which is produced at a very high temperature and in reducing conditions, so that the ferric iron is reduced to a ferrous state, and combines with silica to form a vitreous material that infills the pore spaces in the brick.

The **Mercia Mudstone Group** (formerly 'Keuper Marl') outcrop provided an abundant source of brick clay to the south-east of the Birmingham Fault, but workings ceased in about 1988. The main pits were: Holly Lane, Erdington [120 914], Kings Norton Brick and Tile Works [054 786], Millpool Hill Brickworks [0785 7945], Saltley–Adderley Park [102 874], the Bordesley area [09 86], and Jackson's Brick Pit (Arden Works) [205 830]; the latter was the last working pit (Plate 2). In addition, the Mercia Mudstone was formerly worked in scattered, small pits over much of the drift-free outcrop for **agricultural treatment** ('marling').

The following were formerly worked as minor brick clay resources. Clays of the **Alveley Member** (formerly 'Keele Formation') were locally exploited in small pits at [0213 9055] and [0187 9126], but the clay is generally unsuitable for brick manufacture because of the relatively high calcium carbonate content and low iron content. This unit was also utilised by mixing with overlying **glaciolacustrine clay** at California, near Harborne [036 834]. Glaciolacustrine Clay was also exploited for brick manufacture at Baggots Bridge [971 857], near Moxley, and at Brickfield Farm [195 864]. Clay **till** (Boulder Clay) was locally exploited for brick manufacture at Solihull Lodge [089 786] and [094 787]. **Coal Measures** mudstones, often a local by-product of ironstone workings, were also worked.

Ironstone

Ironstone occurs in the form of nodules or thin beds of siderite (iron carbonate) within Coal Measures mudstone. In the Sandwell area, the Ten Foot Measures, Pouncil and Gubbin Measures were worked along with the Thick Coal. The Poor Robins, Gubbin and Balls, and Blue Flats ironstones were worked from Bromwich Hall Colliery [0026 9443]; the Whitestone (New Mine Stone) and the Brooch Stone were also worked locally. The Gubbin Ironstone (above the Heathen Coal) was the most extensively worked seam, yielding up to 40% metallic iron (Eastwood et al., 1925), but production ceased in the early 20th century.

Coal

Coal and associated **fireclay** and ironstone seams (Chapter Five) have been worked in the South Staffordshire coalfield since medieval times, although the most recent underground workings at Sandwell Park, Heath Pits and Hamstead collieries, located to the east of the Eastern Boundary (Great Barr) Fault, ceased in the early 1960's (Powell et al., 1992) (Figure 3). Coal was also won

in recent years, at depth, along the western margin of the Warwickshire coalfield from Dexter, Kingsbury and Daw Mill collieries (for details see Bridge et al., 1998) (Figure 2). Constraints to development associated with areas undermined are discussed in the section on Foundation Conditions.

In the **South Staffordshire coalfield** the Thick Coal (6–12 m thick) was worked at depths of between 386 and 572 m from Sandwell Park [0199 8980], Hamstead [0431 9296] and Heath Pits [008 899] collieries (Figure 3); other worked seams were the Brooch (1.2–2 m thick) and the Heathen (1.4 m thick). Workings in the Thick Coal from these collieries were limited to the east of Heath Pits and the west of Sandwell Park collieries, by faults bounding the Silurian 'highs' (Waters et al., 1994), and the thinning of the coals towards and over the highs. The Thick Coal was worked at depths of about 580 m below OD in the graben between the Park Farm and Newton faults (Figure 30).

To the west of the Eastern Boundary (Great Barr) Fault, the seams crop out, or are found at relatively shallow depths, below bedrock or superficial deposits. The outcrop of the Coal Measures and overlying Etruria Formation has a high concentration of shafts (Powell et al. 1992) exploiting the main seams, namely, the Thick, Brooch, Heathen, Rubble and New Mine coals. Coal was worked at depths in excess of 150 m below the Rowley Regis intrusion.

Opencast mining ('openworkings') was carried out by the early miners, where the seams, particularly the Thick Coal, came to crop, such as at two small sites at Old Park [983 960] and King's Hill [984 956]. However, the largest opencast working in the district was at the former (1989–90) Patent Shaft Steel Works site [978 950] (Figure 3, Plate 4)), which worked the Rubble, Heathen, Thick and Flying Reed seams, to the north of the Coseley–Wednesbury Fault, and the Brooch, Two Foot, and an un-named coal, to the south of this fault. The Thick Coal was also worked by opencast methods at Mesty Croft [9945 9500] in 1971. Controlled excavation and backfilling, together with remedial work on extant mine shafts has allowed these sites to be redeveloped subsequent to coal extraction. Although coal seams lie at relatively shallow depths in the north-west of the district, the resource potential for opencast coal is limited by both the great extent of former shallow workings and urban sterilisation of the outcrop.

Limestone

Limestone was exploited for use as a flux in the iron industry, and was calcined for agricultural lime and cement. Limestone was formerly worked from Silurian strata, namely in descending sequence, the Much Wenlock (Dudley) Limestone and the Barr Limestone.

The most important limestone resource was the **Much Wenlock Limestone;** it does not crop out in the district but was worked in deep mines (Figure 3), often as extensions from existing coal and ironstone workings, during the 19th and 20th centuries (Ove Arup and Partners, 1983). Constraints to development associated with areas

Plate 2a
Jackson's Brick
Pit (Arden
Works) [205 830],
south face of
quarry (1965)
(GS 706).

A upper part
(about 18 m) of
section (Figure 23)
with cyclic nature
emphasised by
differential erosion
of softer and
harder beds
B feature formed
by thin sandstone
(s; Figure 23) and
hard, cream-
weathering clays
(7; Figure 23).
Triassic, Mercia
Mudstone Group.

a.

Plate 2b East
face of quarry
(1965) showing
cyclic sequence
in upper 28 m
of section
(Figure 23).
Triassic, Mercia
Mudstone
Group
(GS 707).

b.

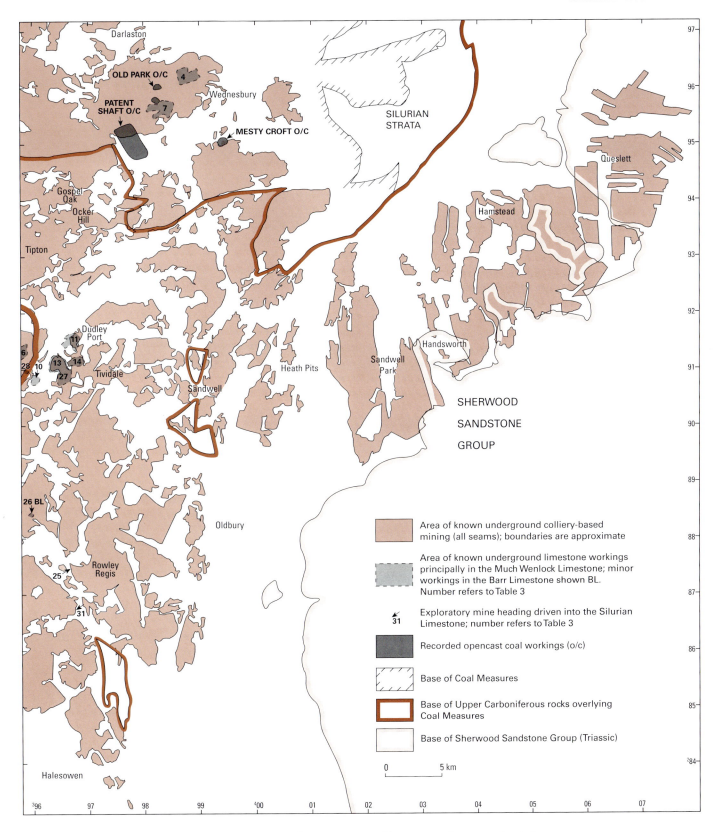

Figure 3 Generalised map showing the approximate extent of known underground mining and opencast coal workings in the west of the district. Based on data from the Coal Authority and Ove Arup and Partners (1993).

Table 3 Silurian limestone mines formerly worked in the Birmingham district; mines index and additional data derived from Ove Arup and Partners (1983) and Powell et al. (1992).

* Exploratory heading driven into Silurian limestone from coal workings
† U Upper Quarried Limestone; Much Wenlock Limestone Formation
L Lower Quarried Limestone; Much Wenlock Limestone Formation
BL Barr Limestone

Index number	NGR (approximate centre of site)	Mine name	Seams worked†	Approximate depth (m) of seam
4	988 962	Blackham	U & L	120 & 165
6	957 913	Coneygre	U	200
7	983 956	Cow Pasture	U	155
10	960 908	Dudley Port, Ensells	U	175
11	966 915	Dudley Port, D & B	U	225
13	964 910	Dudley Port, Giles	U	190
14	967 910	Grovelands	U	195
25*	965 874	Pennant Hill	U & L	125
26	958 883	Springhouse	BL	335
27	965 909	Tividale	U	190
28	959 909	Trough Pits	U	195
31*	967 867	Yew Tree	U & L	220 & 260

undermined are discussed in the section on Foundation Conditions. The formation is subdivided into three members (Chapter Four), but only the Upper Quarried Limestone (about 10 m thick) and Lower Quarried Limestone (about 13 m thick) were worked. Details of the mines are shown in Table 3.

The **Barr Limestone** crops out near Park Hill [037 972] in the north-west of the district and was extensively quarried, in the adjacent Lichfield district, during the late 18th and 19th century, for use as agricultural and builders lime (Price, 1970); exploratory headings into the limestone were driven from adjacent coal workings in the Springfield area (Figure 3, Table 3). Upper Carboniferous *Spirorbis* **limestone** was also worked for agricultural purposes; small pits were dug along the outcrop of the Flanders Hall and Index limestones (Halesowen Formation), and Whitacre and Maxstoke limestones (Meriden Formation) in the east of the district.

Building stone

The **Enville Member (Salop Formation)** was formerly quarried for sandstone at Hamstead Quarry [050 928] and in small pits located to the south-west [0523 9259] and [0578 9226]. Numerous small quarries formerly exploited sandstones of the **Halesowen Formation** and **Meriden Formation** for local building stone, in the South Staffordshire and Warwickshire coalfields, respectively. Although the Triassic **Bromsgrove Sandstone** was widely used for building stone in central Birmingham, it was generally sourced from outside of the district. The most common Triassic sandstones used for municipal buildings in Birmingham city were the Hollington Stone (red, mottled and white varieties), from quarries at Hollington, near Tean, Stoke-on-Trent, and the Grinshill Stone (yellow, creamy white and red varieties) which was worked at Grinshill, Yorton near Shewsbury (Fawdry, 1913; Ashurst and Dimes, 1990).

WATER RESOURCES

Most of the potable public water to the Birmingham area is supplied from Wales via the Elan Aqueduct. Ground-water has been an important source of supply in the past, and is still exploited chiefly for industrial purposes and, to a lesser extent, for agricultural purposes. The principal groundwater source is the Sherwood Sandstone aquifer which, as an unconfined and confined aquifer, underlies much of the city of Birmingham, and was formerly pumped extensively to supply the industries in the city.

Assessment of groundwater is important in the siting of waste disposal sites, and industries likely to exploit, or possibly pollute, the aquifer. Additional information on wells and groundwater in the district is available from the BGS.

Most of the district lies within the Trent River basin. The River Tame drains the northern part of the Black Country (mostly Carboniferous and Permian rocks), crossing the outcrop of the Sherwood Sandstone Group between Hamstead and Nechells to flow eastward and northward towards the River Trent. Farther south, the Sherwood Sandstone Group outcrop is drained by the eastward flowing Bourn Brook, a tributary of the River Rea which in turn, drains the southern outcrop of the Mercia Mudstone Group. The Mercia Mudstone outcrop is largely drained by the northward flowing River Cole and its tributaries, and by the River Blythe; the confluence of these rivers with the Tame is located near Hams Hall, from whence the Tame flows northward. A major drainage divide is located close to the south-western corner of the district, beyond which is the catchment of the River Stour. The drainage pattern has been modified by an extensive system of canals which is centred upon the industrial regions of Birmingham and the 'Black Country'. Average annual rainfall is more than 800 mm in the south-west of the district, but only 650 mm in the north-east. Mean annual evaporation is about 580 mm.

The hydrogeology of the Trent River basin, of which the Birmingham district forms the westernmost part, has been described by Downing et al (1970). The main hydrogeological units are bounded by major faults; the Western Boundary Fault of the Warwickshire Coalfield, the Eastern Boundary (Great Barr) Fault of the South Staffordshire Coalfield, and the north-east trending Birmingham Fault that bisects the area (Figure 4). Numerous spring zones occur along the main fault zones

Figure 4
Hydrogeology of
the district.

a. Distribution of
hydrogeological
units and location
of springs,
groundwater
hydrograph
boreholes and
groundwater
hydrochemistry
sites.

Numbers 1–9 see
Table 5
A, B see Figure 5
C, D, E see text

b. Distribution
of licensed
groundwater
abstraction sites
within the district
(1995).

c. Distribution of
borehole specific
capacities within
the district.

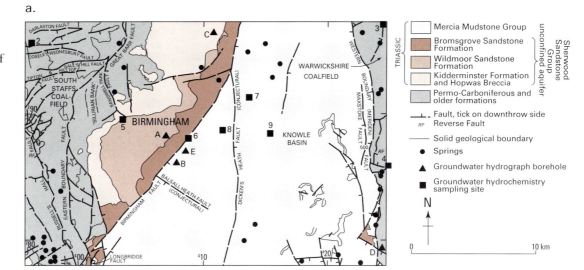

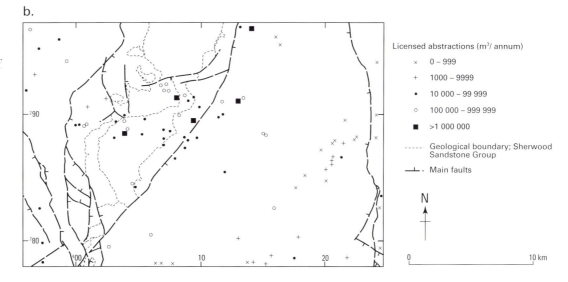

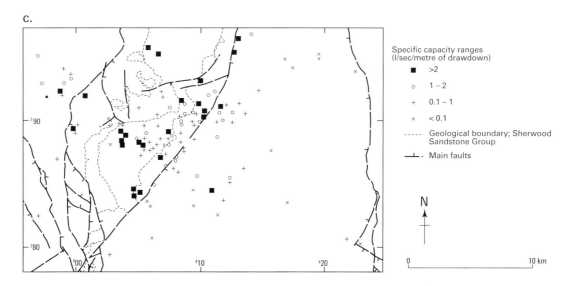

Table 4   Licensed abstractions in the Birmingham district (1994–95). Compiled from data supplied by the former National Rivers Authority, Severn–Trent Region.

Abstractor type	Aquifer			
	Permo–Carboniferous	Sherwood Sandstone	Mercia Mudstone	Totals
Public water supply	— (—)	1 095 000 (1)	— (—)	1 095 000 (1)
Agriculture	125 152 (11)	— (—)	34 495 (22)	161 647) (33)
Industry	1 571 200 (14)	14 741 760 (44)	12 502 (1)	16 325 462 (59)
Domestic	145 000 (3)	— (—)	— (—)	145 000 (3)
Totals	1 841 352 (28)	15 836 760 (45)	48 997 (23)	**17 727 109** **(96)**

Explanation of figures:
m^3/annum
Number of licences shown in brackets

and along marl/sandstone (skerry) boundaries within the Mercia Mudstone Group (Figure 4a). The distribution of boreholes with higher yields and specific capacities tends to mirror the distribution of fault zones within the area (Figure 4b, c).

Limited quantities of potable water are obtained from fractured Permian and Carboniferous strata, primarily for industrial use within the South Staffordshire and Warwickshire coalfield basins.

Land (1966) described the hydrogeology of the Sherwood Sandstone Group within the Birmingham area, where it forms the major unconfined aquifer. To the south-east of the Birmingham Fault the Sherwood Sandstone aquifer is confined by the Mercia Mudstone Group (Figure 4a). This aquifer was formerly much used as a source of water for industrial and domestic purposes, but it has been polluted, locally, with industrial chemicals. The decline of industrial activity within the Birmingham area, and consequent reduction in water demand, has resulted in a marked rise in water levels within the Sherwood Sandstone aquifer system (Knipe et al., 1993).

Groundwater abstraction licence data for the Sherwood Sandstone aquifer and other minor aquifers, including water use, are shown in Table 4. The location of licensed abstraction boreholes are shown on Figure 4b. Currently, water abstracted from the Sherwood Sandstone aquifer is used mostly for industrial purposes. Formerly, large-scale abstraction of groundwater was undertaken to meet the needs of a variety of industries and for public water supply. The industrial needs have been reduced or removed with the decline of water intensive industry within the West Midlands area. In addition, much of the water requirement of the West Midlands conurbation is now met from central Wales via the Elan aqueduct. Abstraction for public supply is undertaken by the South Staffordshire Water Company at Perry Barr [1380 9657] at a rate of 1095 megalitres per annum (Mla^{-1}).

The Mercia Mudstone Group is generally regarded as impermeable. However, small quantities of potable water are obtained from the Arden Sandstone and other thin sandstones (skerries) interbedded within the group. The large number of licences are indicative of the importance of these minor aquifers, particularly to the agricultural economy of the area.

Recharge to these aquifer systems, especially the unconfined Sherwood Sandstone aquifer of the Birmingham area, is largely controlled by the distribution and lithology of the superficial (drift) deposits, which consist predominantly of till, sand and gravel. Infiltration and groundwater flow is also influenced by the presence of several drift-filled, buried channels (palaeovalleys) incised into the bedrock.

Hydrogeological characteristics of the principal aquifers

PRE-CARBONIFEROUS ROCKS

Small quantities of groundwater occur in fracture zones within the Lickey Quartzite (Ordovician) and Silurian limestones and shales. In the west of the district, around Wednesbury [98 95] and Dudley Port [96 91], the Silurian strata locally contain small quantities of groundwater, particularly within abandoned limestone mines (Much Wenlock Limestone).

CARBONIFEROUS AND PERMIAN ROCKS

Carboniferous and Permian strata crop out in the Warwickshire Coalfield and the South Staffordshire Coalfield (north-west Birmingham), to the west of the Eastern Boundary Fault (Figure 1). Highly fractured **Coal Measures** (about 150 m thick) occur within both coalfields. In both areas these heterolithic strata form complex multi-layered aquifers, with alternations of thick

Table 5 Typical chemical analyses of groundwater from selected boreholes in the district.
Data supplied by the (former) National Rivers Authority, Severn–Trent Region.

Location and number (see Figure 4a)	1	2	3	4	5	6	7	8	9
	Bass Brewing	British Waterways	Heanley Farm	Warren Farm	Yorkshire Imp. Metals	Central Midland Co-op	Jaguar Cars	Southalls	Alcan Plate
NGR	SP 0170 9770	SO 956 950	SP 2500 9680	SP 2510 8490	SP 0290 8880	SP 0840 8730	SP 1290 9060	SP 109 877	SP 1486 8768
Aquifer*	Permo-Carb.	Permo-Carb.	Permo-Carb.	Permo-Carb.	SSG	SSG	SSG	SSG	SSG
Date of analysis	24/8/94	24/8/94	23/3/94	23/6/93	4/10/93	9/11/93	30/6/93	27/2/92	30/6/93
pH	7.4	6.9	7.2	7.3	7.5	7.7	8.0	7.6	7.7
Conductance(μs/cm)	970	1720	1070	680	1050	520	320	700	810
Ca^{2+} mg/l	112	193	215	142	149	44.7	35.8	81.6	7.3
Mg^{2+} mg/l	39.7	84.6	9.68	4.14	37.1	32.5	9.6	48.7	29.1
Na^+ mg/l	60.7	113.0	19.4	7.3	43.0	21.5	19.0	23	68.0
K^+ mg/l	3.2	21.0	40.0	2.4	4.7	4.1	1.4	2.6	3.2
HCO_3^- mg/l	290	491	292	216	253	188	104	156	117
SO_4^{2-} mg/l	175	379	122	63.8	195	53.5	62.9	297	296
Cl^- mg/l	70	124	59.1	25.1	68.8	32	5.26	13	11.2
Total hardness mg/l	443	830	576	372	526	246	129	407	310

* Aquifers: Permo-Carb. = Permian to Carboniferous strata
  SSG = Sherwood Sandstone Group

'shales' and thin sandstones. The 'shales' act as aquitards or aquicludes, with the sandstones forming fractured aquifers of limited lateral extent. The development potential of these sandstone aquifers, where secondary permeability predominates, is usually limited. Closure of mines and decreased rates of pumping is causing rising water levels, but no data are yet available to indicate the magnitude of the problem within the coalfields. Few data are available regarding groundwater levels within the Coal Measures and the effects of land subsidence upon aquifer conditions.

Within the South Staffordshire Coalfield yields of 1 to 12 litres per second per metre (ls^{-1}) have been produced from boreholes drilled into the Coal Measures, notably in the West Bromwich area, where, exceptionally, one borehole (West Midlands Gas Board [9903 9246]) produced 64 ls^{-1}. Several high yielding boreholes with licensed abstraction rates of 300–500 Mla^{-1} from the Coal Measures are located within the South Staffordshire area. Borehole specific capacities of 0.1 to 2 litres per second per metre ($ls^{-1}m^{-1}$) of drawdown are obtained within the central parts of the coalfield, increasing to > 2 $ls^{-1}m^{-1}$ of drawdown adjacent to the Eastern Boundary Fault (Figure 4c). Groundwater quality can be very variable, but is generally of calcium bicarbonate type, hard, with low chloride concentrations (generally < 100 mg/l), and high sulphate concentrations (up to 379 mg/l) (Table 5, Locality 2). Elevated sulphate and iron concentrations result from the dissolution of iron pyrites in the Coal Measures. The South Staffordshire Coalfield area, in particular, has a long history of heavy industry resulting in deterioration of groundwater quality in many places, especially where there have been direct discharges of liquid industrial waste into old mine workings. The **Etruria Formation** (30–250 m thick) diachronously overlies the Coal Measures and forms an aquitard composed of mudstone with subordinate pebbly sandstone and conglomerates.

The succeeding **Halesowen Formation** (50–120 m thick) consists of calcareous sandstone and subordinate mudstone with sparse thin limestones and thin coal seams. Where fractured these sandstones can possess significant secondary permeability that result in yields of 5 to 22 ls^{-1}, although the water produced can contain high levels of permanent hardness.

The **Alveley Member (Salop Formation)** (50–275 m thick) of the South Staffordshire Coalfield consists primarily of mudstone with subordinate sandstones, sparse thin limestones and 'cornstones'. Although of low permeability, a yield of 2 ls^{-1} was obtained from the unit at one well [9958 8832], and several small-scale licensed abstractions are current from this member via boreholes located within the Wolverhampton area. The overlying **Enville Member (Salop Formation)** (60–120 m thick) is composed of mudstone, sandstone, cornstones, and calcareous conglomerates, and is generally unconformably overlain by the permeable Kidderminster Formation (Sherwood Sandstone Group). Boreholes in this member have yielded 7.5 to 10 ls^{-1}.

In the Warwickshire Coalfield the Upper Carboniferous red-bed strata comprise the **Meriden** and **Tile Hill Mudstone** formations, include one of the main aquifers in the Coventry area (Lerner et al., 1993). Within the water-bearing strata the higher the ratio of sandstone to interbedded marl, the greater will be the degree of secondary permeability derived from fracturing. The **Corley Sandstone** which forms the upper part of the **Keresley Member** (Meriden Formation) is the main water-bearing horizon within the western part of the Warwickshire Coalfield area; several boreholes in this area have yielded 0.4 to 0.6 ls^{-1}. Currently there are six boreholes abstracting from the Upper Carboniferous strata in this area; four with licensed abstraction rates of 0.7 to 0.8 Mla^{-1} , and two with rates of 45 to 61 Mla^{-1}.

The Carboniferous **dolerite** at Rowley Regis is regarded as being impermeable, although there may be some fissure flow in the highly fractured and highly jointed rock near the margin of the intrusion.

PERMIAN AND LOWER TRIASSIC ROCKS

The Permian **Clent Formation** (up to c.113 m thick), penetrated in water boreholes in central Birmingham, is composed of red marl, sandstone and breccia, and in this area is generally overlain by the Triassic **Hopwas Breccia** (c.10 m thick) which is laterally discontinuous and is composed of coarse pebbly sandstone and quartzite breccia. These units are in hydraulic continuity with the overlying Kidderminster Sandstone Formation (Sherwood Sandstone Group aquifer) and have produced high yields in the Nechells area.

TRIASSIC SHERWOOD SANDSTONE GROUP AQUIFER

The Kidderminster Formation, Wildmoor Sandstone Formation and Bromsgrove Sandstone Formation are each important sandstone aquifers, but because of the limited amount of reliable hydraulic data for the specific formations they have been grouped, together with the underlying Hopwas Breccia, as the Sherwood Sandstone Group aquifer. The nature of the three arenaceous formations that comprise the Sherwood Sandstone Group within the Birmingham area has been studied in detail by several workers (Wills, 1976 and references therein).

The Sherwood Sandstone Group represents the main aquifer of the Birmingham district from which large yields, in excess of 4 million litres/day, were obtained from shafts up to 3 m in diameter and boreholes more than 600 mm in diameter (Lovelock, 1977). The depths of the shafts are normally less than 100 m, whilst the depths of the boreholes are greater than 200 m, particularly within the confined zone of the aquifer. The thickness of the saturated zone was commonly more than 60 m. Within the confined zone fully flowing artesian boreholes were rare. The mean annual infiltration into the sandstone outcrop is estimated to be about 170 mm per annum (Monkhouse, 1986).

The **Kidderminster Formation** comprises 45 to 120 m of predominantly medium- to coarse-grained, pebbly sandstone with poorly cemented conglomerate towards the base. It varies from friable and weakly cemented to well cemented. It has a median porosity of 29% and hydraulic conductivities of 3.8 to 7.7 metres per day (m/d). The median horizontal permeability (Kh) value is 3.5 m/d with a median vertical permeability (Kv) value of 2.7 m/d. Local cementation may reduce these figures to around 20% porosity and hydraulic conductivities of 0.6 m/d (Lovelock, 1977). The unconsolidated conglomerates also have good aquifer properties. An intergranular transmissivity (Ti) of 70 m^2/day is likely, assuming an aquifer thickness of 20 m. Lovelock (1977) estimated the Ti value to be 268 m^2/day, while the total transmissivity, T, was found to be 2771 m^2/day, indicating fracture flow to be dominant.

Numerous boreholes yielding in excess of 20 ls^{-1} have been sunk in this formation, mainly at gas works and brewery sites. Several deep (> 300 m) boreholes drilled through this formation into the underlying Hopwas Breccia and Clent Formation within the vicinity of fracture zones have produced very high yields, for example: the Birmingham Corporation Water Works boreholes at Aston [0916 9035] (182 ls^{-1}), Short Heath [0940 9260] (104 ls^{-1}), and Perry Well [0771 9217] (104 ls^{-1}). High specific capacities in excess of 2 ls^{-1}m^{-1} of drawdown have been recorded from numerous high yielding boreholes drilled through this formation (Figure 4c). The groundwater is typically of the calcium carbonate type; chloride concentrations are generally less than 30 mg/l.

The **Wildmoor Sandstone Formation** crops out under much of west Birmingham, but has a smaller areal extent and, in general, cannot be traced as a distinct unit to the southeast of the Birmingham Fault. It underlies a large part of the city and consists of up to 120 m of fine- to medium-grained, thick bedded, soft, sandstone with partings and thin beds of marl. Lovelock (1977) reported hydraulic conductivities of between 0.06 m/day and 0.6 m/day from limited data (with a median Kh of 1 m/day and median Kv of 0.83 m/day with a median porosity of 27%. The value for Ti was 100 m^2/day, assuming an aquifer thickness of 100 m.

The **Bromsgrove Sandstone Formation** is the upper subdivision of the Sherwood Sandstone Group. It forms a narrow outcrop to the north-west of the Birmingham Fault, but is concealed beneath the Mercia Mudstone Group to the south-east, on the downthrow side of the fault where it attains a maximum thickness of 180 m or more. The formation consists primarily of well-cemented sandstone, pebbly sandstone and conglomerate with beds of mudstone, locally up to several metres thick. The effect of this lithological inhomogeneity upon the hydraulic characteristics of the Bromsgrove Sandstone aquifer has been studied by Rushton and Salmon (1993). The coarsest sandstones have hydraulic conductivities of the order of 6 m/day with porosity varying from 23 to 31% depending on the degree of cementation (Lovelock 1977). The medium-grained sandstones tend to be clean and well-sorted, with hydraulic conductivities usually ranging from 1.3 to 6 m/day and with porosity of around 30%. Fine-grained sandstones range from unconsolidated to well cemented; the commonest are consolidated slightly silty sandstones, with hydraulic conductivities in the range of 0.26 to 0.39 m/day and porosity of 27 to 30%. Median values for this sequence are Kh of 0.93 m/day, Kv of 0.53 m/day and a porosity of 28%; Ti is 47 m^2/day assuming an aquifer thickness of 50 m. In the upper part of the Bromsgrove Sandstone Formation there is a gradual change in lithology from sandstone to mudstone of the overlying Mercia Mudstone Group. These passage beds consist of intercalations of hard silty, micaceous sandstone with an upward increasing proportion of mudstone beds.

To the south-east of the Birmingham Fault the aquifer is overlain and confined by up to 300 m of Mercia Mudstone Group in which mudstone is predominant. Numerous deep boreholes have been drilled into the confined aquifer within the area between the Birmingham and Dicken's Heath faults. Yields of 10 to 40 ls^{-1} with specific capacities of 0.1 to 2 la^{-1}m^{-1} of drawdown have been determined from the confined aquifer. Good quality water of calcium bicarbonate or calcium sulphate types have been obtained. Sodium and sulphate concentrations increase to the south east, away from the fault zones (Figure 4a; Table 5).

MERCIA MUDSTONE GROUP

The Mercia Mudstone Group consists of up to 365 m of mudstone and occasional thin siltstone/sandstones beds ('skerries') which are more common in the lower part of the unit; gypsum (calcium sulphate) occurs as thin beds and nodules throughout much of the group, but is generally absent within 30 m of the ground surface as a result of dissolution in this zone. Overall, the Mercia Mudstone Group is regarded as weakly permeable. Small quantities of groundwater (Table 4) may occur within the skerries.

The **Arden Sandstone Formation** (10 m thick), occurs about 30 to 40 m below the top of the Group and is the source of numerous springs in the south-east of the area (Figure 4a). Small yields of 0.3 to 1 ls^{-1} of reasonable quality groundwater (Table 4) have been obtained primarily from the Arden Sandstone Formation at numerous points within the Knowle Basin (Figure 4b). These restricted water resources are important to the agricultural communities of the area. Water quality is generally potable although very hard.

SUPERFICIAL DEPOSITS

The thickness and type of glacial drift covering the Sherwood Sandstone aquifer of the Birmingham area control the amount, and rate, of infiltration of precipitation into the aquifer. Horton (1974) distinguished four broad lithological types.

1. Till with low permeability.

2. Glaciolacustrine deposits consisting of clay, silt, fine-grained sand and peat, with poor vertical permeability, but better horizontal permeability.

3. Glaciofluvial sand and gravel with relatively high permeability.

4. Interglacial deposits (humic silt and clay), of low permeability.

Clayey deposits (till) predominate within the south-western part of the area, west of a line through eastern Smethwick, Rotton Park Reservoir, Harborne and Selly Oak, whereas more sandy drift deposits blanket much of the area to the east. Areas of sandy drift would be expected to permit more rapid and direct infiltration of precipitation or seepage from water courses, water mains and sewers. Infiltrating water will gravitate down to the sandstone bedrock via a route made more circuitous by interbedded clay or till lenses.

The incised proto-Tame, Moxley, proto-Ford, Millfield and Hockley channels (Chapter Eight) are important to subsurface groundwater flows in the Birmingham area since the predominantly sandy fill permits hydraulic continuity with the bedrock. The Hockley Channel is a steep-sided buried channel, up to 400 m wide and 35 m deep infilled with gravels and sandy clays. In the Aston area the base of the channel is at a level of 65 to 70 m above OD, some 30 to 35 m below the level of the present River Rea. The floor of the palaeovalley rises before it reaches the Birmingham Fault; there may be limited flow between the alluvial/glacial sequence and the Sherwood Sandstone Group aquifer.

FLUCTUATING GROUNDWATER LEVELS

Rising water table within the Birmingham area

Knipe et al. (1993) reviewed the development of groundwater abstraction within the Birmingham area for the period from 1850 to 1992, and have tried to predict the effects of groundwater rise during the next 20 years. Until the 1960's, heavy industrial use of groundwater led to a drawdown of the water table which reversed when abstraction gradually ceased. This rebound is predicted to have an important effect on foundation conditions, seepage into excavations and remobilisation of potentially harmful substances.

Prior to the 1850's groundwater appears to have discharged into the rivers Tame and Rea from Triassic sandstones to the north-west of the Birmingham Fault via a series of spring zones occurring along the length of that fault. Little or no flow occurred across the fault to the south-east due to the presence of relatively impervious Mercia Mudstone on the downthrow side of the fault. Much of the low-lying Nechells area was, at that time a wetland area. The postulated pre-abstraction groundwater piezometer levels are shown in Figure 5a.

By 1855, large quantities of groundwater were being abstracted from the Sherwood Sandstone Group aquifer, both from the unconfined aquifer north-west of the fault and from the confined aquifer to the south-east of the fault. Maximum levels of abstraction were achieved at the time of the Second World War. Rates of water abstraction dropped with the post war decline in manufacturing industries in the Birmingham area, the highest rate of decrease being during the late 1960s. The groundwater

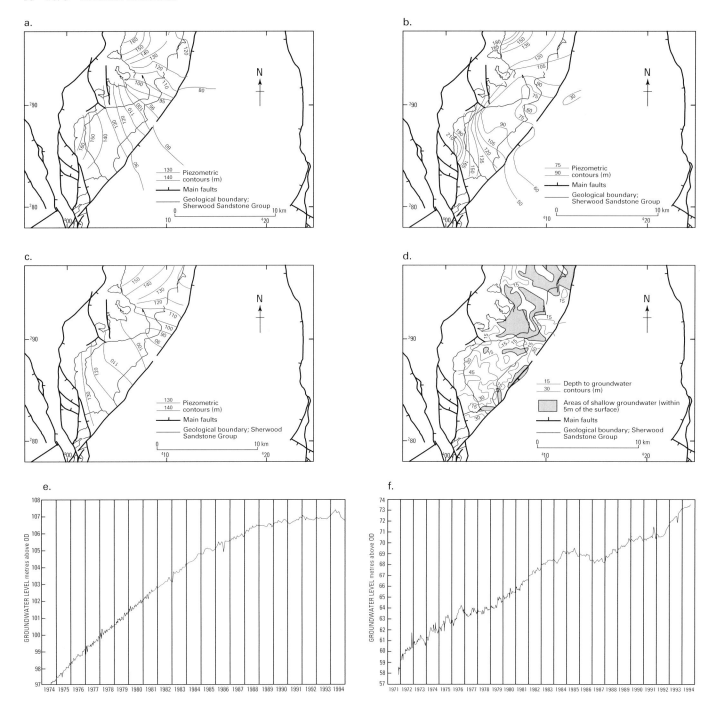

Figure 5   Groundwater fluctuation in the district.

a. Postulated pre-abstraction groundwater piezometric contour map (after Knipe et al., 1993).
b. Groundwater piezometric contour map, 1966 (after Land, 1966).
c. Groundwater piezometric contour map, 1988-89 (after Knipe et al., 1993).
d. Estimated depth to groundwater, year 2020 (after Knipe et al., 1993).
e. Groundwater hydrograph for the Constitution Hill Borehole [SP 0670 8759] (site A on Figure 4a).
f. Groundwater hydrograph for Dares Brewery Borehole [SP 0761 8516] (site B on Figure 4a).
    Hydrographs provided by the (former) National Rivers Authority, Severn–Trent Region.

piezometric contour map for 1966 (Land, 1966) indicates declines in water table levels of the order of 30 m within central Birmingham (Figure 5b). By the 1988–89 period the much reduced levels of abstraction had resulted in a recovery of water table levels by as much as 20 m in large areas of central Birmingham (Figure 5c).

Two hydrographs from boreholes located within central Birmingham illustrate the ongoing rates of water level rise. Hydrograph 'A' (Figure 5e) indicates that within the Constitution Hill area [0670 8759] a fairly constant rate of recovery of about 0.7 m/annum was sustained from 1974 to 1982. Since then the rate has declined to about 0.2 m/annum. At the latter rate the effect of climatic events, such as the 1992 drought, begin to cause fluctuations on the graph. To the south-east of the Birmingham Fault a rate of recovery of 1 m/annum was recorded of hydrograph 'B' at the Dares Brewery Borehole [0761 8516] until 1984 (Figure 5f). Since 1987, following a short period of abstraction, the recovery rate has been sustained at 0.7 m/annum. The effects of the 1992 drought are also apparent on this hydrograph. At an adjacent borehole at Upper Trinity [0834 8623] hydrograph 'E' recorded a rate of recovery of 0.3 m/annum during 1971–1982, since when a much reduced rate of 0.1 m/annum has been sustained. Control hydrographs located at Sutton Park and Ram Hall observation boreholes (Figure 4a; C and D, respectively) located outside the zone of former depletion, show only expected annual variations in water table levels due to seasonal and major climatic events such as the long-term drought that culminated in 1992 which caused a 2 m decline in water table at Sutton Park and a 3 m decline at Ram Hall.

Knipe et al. (1993) predict that, by the year 2020, groundwater levels should lie within 5 m of the land surface within the low-lying areas located along the line of the Birmingham Fault and along the Tame valley within the north-eastern parts of Birmingham (Figure 5d). Such water levels would have detrimental effects upon building foundations, sewerage and drainage systems and construction of major civil engineering works such as the Birmingham Metro System. The flooding of low-lying areas and inundation of building foundations may increase in frequency as water levels continue to rise toward their pre-development levels. Rising groundwater in low-lying industrial areas along the valleys of the Tame, Rea and Hockley Brook, where the natural ground or fill materials are chemically contaminated, could result in leaching and remobilisation of contaminants into the groundwater, and subsequently into the water courses.

Areas of shallow groundwater in the city generally coincide with areas of made ground/backfilled ground or other weak soil conditions (Knipe et al., 1993). South of the River Tame, groundwater rises are estimated to be significant, whereas north of the river Tame little fresh damage is predicted (Knipe et al., 1993). However, around Wilton, adjacent to the M6 motorway, some properties have already suffered inundation of basements, and subsidence damage possibly due to wetting of the

subsoil. In the Birmingham city area deep tunnels, such as those for the Metro and for telecommunications, are already affected by rising groundwater (Knipe et al., 1993). Whilst it is the case that a rise in moisture content generally reduces the compressive strength of sandstone, structures founded on sandstone bedrock are unlikely to be affected by groundwater rise. However, this is not the case for drift materials, artificial deposits (made ground and fill) and Mercia Mudstone. The bearing capacity of saturated silty clayey sand and gravel, as compared with unsaturated, could be reduced by as much as 50% (Knipe et al., 1993). It may be that, given sufficient groundwater rise, instability of manmade or even natural slopes may be promoted in superficial sediments overlying the aquifer.

AQUIFER VULNERABILITY AND POLLUTION

The hydrochemical studies of the groundwaters from the Sherwood Sandstone aquifer in the Birmingham area (Edmunds et al., 1989) indicate that the waters were all of calcium bicarbonate type with high sulphate contents; high Cr and Br contents were also noted. Mg/Ca ratios are less than 1 implying that bicarbonate is the major mineral giving rise to mineralisation.

Hydrochemical studies of the Sherwood Sandstone aquifer in the vicinity of the Birmingham Fault (Jackson and Lloyd, 1983) concluded that the combined effects of faulting and the irregularity of the base of the Sherwood Sandstone aquifer are major controls on the type of water flowing across the fault. Faulting within the confined section has been shown to be a major influence upon the location of high-sulphate groundwaters.

Studies of groundwater contamination by chlorinated solvents and other pollutants within the urban, industrialised areas in central Birmingham (Rivett et al., 1990; Ford et al., 1992), identified three main contaminants:

TCE trichloroethane — used for metal cleaning (over 90% of UK use), dry cleaning and extractions. Introduced in 1910 but used widely from 1928 onwards until replaced by TCA from 1965 onwards.

TCA 1,1,1 trichloroethane — used for metals/plastics cleaning, adhesives, aerosols, inks etc.

PCE tetrachloroethane — used for dry cleaning, metal cleaning, intermediates in processes.

Some contamination of the aquifer has resulted from casual use and indiscriminate disposal of solvents. The greatest solvent contamination was observed where the drift deposits are thin or permeable. However, some confined aquifer boreholes did contain high concentrations, for example 97 mg/l under 45 m of Mercia Mudstone. In such cases solvents are likely to have drained down the outside of a poorly completed borehole, or have been illicitly dumped into a disused borehole.

The contamination of the Birmingham Sherwood Sandstone Group aquifer is largely due to TCE and to a lesser extent TCA, both of which have been widely used in metal cleaning. Some PCE contamination has arisen from dry cleaning. The stability of TCE concentrations observed in the Birmingham aquifer suggests that contamination would persist for many decades even if use of the solvent were discontinued. The majority of borehole contamination incidents in the Birmingham survey could be ascribed to sources of solvents local to the boreholes. Thick, low-permeability drift deposits or bedrock help, but do not guarantee, to protect underlying groundwater. Monitoring of boreholes over many months revealed that solvent concentrations were often reasonably constant with time. When major variations did occur they could be accounted for by either variation in the abstraction history of the borehole, or migration of solvent plumes during monitoring.

Ford et al. (1992) also concluded the Birmingham aquifer is contaminated by TCE. The occurrence of chlorinated solvents can be related to hydrogeological factors. The overlying strata, thickness of unsaturated zone, depth of solid borehole casing and abstraction history have important influences on solvent levels.

In an effort to minimise possible pollution of the main aquifers from contaminants dumped in landfill sites, such points of waste disposal are now preferentially located upon impermeable strata such as the Mercia Mudstone Group, or upon strata known to contain poor quality or already contaminated groundwater such as the Coal Measures of the South Staffordshire Coalfield. Waste disposal sites within the West Midlands are mostly restricted to lower risk sites. Unfortunately, the same cannot be said for the distribution of waste transfer sites and scrap vehicle dealers, many of which are located upon the unconfined Sherwood Sandstone Aquifer. Spillage of contaminants and petrochemicals at such sites can only lead to further contamination of the Sherwood Sandstone Aquifer system.

FOUNDATION CONDITIONS

The suitability of bedrock and superficial materials for foundations depends mainly on their geotechnical properties. The various parameters relating to the bearing strength and the reaction of engineering structures to bedrock, superficial (drift) and artificial deposits are reviewed, below. In addition, the stability of engineering structures is also related to geological factors such as local geological structure (Chapter Nine), slope stability, the possible presence of mining cavities (see Undermining section) and foundation/tunnelling problems associated with rising groundwater (see previous sections). These factors may give rise to difficult ground conditions which may act as a constraint to development. Although most ground problems can be overcome by appropriate engineering or construction works, the early recognition of problems which may arise is of great importance, because unforseen problems lead to additional costs and delays, and, in some cases solutions may be prohibitively expensive.

Appropriate site investigation (Anon, 1981b) should be carried out prior to development. BGS holds archival geological and topographical data relating to former land use, and both borehole and geotechnical information which are of prime importance in assessing the suitability of a site for development or redevelopment.

The engineering geological classification of the principal bedrock units and drift deposits is shown in Table 6. Descriptive terminology used in the classification of parameters such as strength, consistency and texture of the geological materials is listed in Table 7, and the classification of sulphate content is shown in Table 8. Lithological details of the various units can be found in the relevant chapters.

Bedrock

Geotechnical properties of the principal bedrock units at outcrop are described in ascending stratigraphical sequence (Table 1).

SILURIAN

The Coalbrookdale Formation consists of heavily over-consolidated, fossiliferous, greyish green siltstone and mudstone with thin, nodular limestone beds. Siltstones and mudstones weather to a clay or clayey silt. Joints within the limestone may be clay-filled. Strength data indicate the formation is a 'weak' rock in its unweathered state, and 'firm' to 'very stiff' in its weathered state, with 'low' to 'intermediate' plasticity. However, localised zones may have 'high' plasticity. The sulphate content is typically Class 1 (low).

CARBONIFEROUS AND PERMIAN

The **Coal Measures** consist of mudstone, siltstone, sandstone, with subordinate coal, fireclay, and ironstone, of which about 13–17 per cent by volume is coal. The sulphate content is typically Class 1 (low), but Class 2 also occurs. Rock mass properties are very variable.

No geotechnical data are available for the **dolerite** intrusions at Rowley Regis (Chapter Five) area (Forster, 1991). The rock mass contains abundant vertical and sub-horizontal joints which influence quarrying operations and may have a bearing on the stability of engineered slopes. Low-angle, curved joints at the margin of the intrusion are susceptible to failure.

The Etruria Formation (Etruria Marl) consists of red (and variegated grey, brown, purple, and yellow), hard, fissured mudstone, siltstone, and silty mudstone, with lenses of coarse greenish grey sandstones, grits, and conglomerates ('espleys' or 'espley rock'). The 'espley rock' may be highly weathered. Thin poor quality coal is also present in the lower part of the formation. Geotechnical properties of Etruria 'Marl' at Bury Hill, Rowley Regis have been described in relation to a specific case of slope instability (Hutchinson et al., 1973); at this locality the formation is described as strongly aggregated, with iron compounds as the aggregating agent (aggregation ratio of 2 to 3). Mudstones weather to a clay which ranges from 'soft' to 'very stiff', whilst plasticity is 'low' to 'intermediate' (Forster, 1991). The formation is typically mantled by head, particularly on the steep slopes adjacent to the Rowley Regis intrusion. The formation

Table 6 Classification of likely engineering behaviour of the principal bedrock units and superficial (drift) deposits.

A	SOLID (BEDROCK)	
1	STRONG ROCK	Dolerite
2	SANDSTONE (WITH CONGLOMERATE)	Bromsgrove Sandstone Formation Wildmoor Sandstone Formation Kidderminster Formation (sandstone)
3	3a MUDSTONE (WITH SILTSTONE AND SANDSTONE)	Mercia Mudstone (Zones I & II) Coalbrookdale Fm (Wenlock Shales)
	3b INTERBEDDED MUDSTONE AND SANDSTONE	Clent Formation (breccia-dominated) Salop Formation Tile Hill Mudstone Formation Meriden Formation Halesowen Formation Etruria Formation Coal Measures
B	SUPERFICIAL (DRIFT) DEPOSITS AND WEATHERED MERCIA MUDSTONE	
1	COHESIVE	
	1a OVERCONSOLIDATED	Mercia Mudstone (Zones III and IVa) Till and sandy till
	1b NORMALLY CONSOLIDATED	Mercia Mudstone (Zone IVb) Alluvium Lacustrine deposits
2	NON-COHESIVE	Glaciolacustrine deposits Glaciofluvial terrace deposits Glaciofluvial deposits River Terrace deposits
3	HETEROGENEOUS	Head Pleistocene mass-movement (landslip) Made ground/fill worked ground

may be excavated by digging, but pebbly sandstones may require ripping. It has a good bearing capacity where not subject to softening. The stability of excavations and slopes may require attention, particularly where head or fill are also present. The sulphate content is typically Class 1 (low).

The **Halesowen Formation** consists of grey-green mudstone and siltstone, with thick yellow-weathering, locally pebbly, sandstones, with subordinate thin coals (generally not worked), seatearth fireclays, and *Spirorbis* limestone. Calcretes occur in the mudstones. Clays are of 'low' to 'intermediate' plasticity. The formation may be excavated by digging and ripping, with the possible exception of some sandstone beds. It has a good bearing capacity; the sulphate content is typically Class 1 (low).

The **Alveley Member (Salop Formation)** consists mostly of orange-red, red-brown, fissured mudstone, siltstone, and sandstone, with subordinate *Spirorbis* limestone beds. Sandstones commonly contain pellets or clasts of mudstone and/or caliche; nodular caliche also occurs in the mudstones. The massive sandstone is usually separated into large, discrete blocks by near-vertical joints, which are commonly open, and in some cases may be disrupted by mining subsidence. The member may be excavated by ripping near surface, depending on the thickness and condition of the sandstone. It has a good bearing capacity; the sulphate content is typically Class 1

(low). Sandstones of the Alveley Member may be locally poorly cemented and 'very weak'.

The Enville Member (Salop Formation) and Meriden Formation are heterolithic units generally consisting of sandstone, conglomerate (limestone, dolomite, and chert clasts) and mudstone; *Spirorbis* limestone beds are present at some levels in the Meriden Formation. The sandstones commonly contain clasts of mudstone. Few geotechnical data are available. Clays are generally calcareous and of 'low' to 'intermediate' plasticity. The units may be excavated by digging and ripping near surface, and have good bearing capacity.

The **Clent Formation**, in the south-west of the district, consists of breccia with a 'weak', clayey, locally carbonate-rich matrix, and weathers to a clayey gravel (Powell et al., 1992); pebble- to cobble-grade clasts consist mainly of Uriconian volcanic rocks. Clay plasticity is 'low' to 'intermediate'. Few geotechnical data are available for this formation. The sulphate content is typically Class 1 (low).

TRIASSIC

Sherwood Sandstone Group

The sandstones of this group have a wide range of strength, from massive, well-cemented, 'very strong' to flaggy, weakly cemented, friable 'weak' rock. Sandstones of the Wildmoor Sandstone Formation may be locally poorly cemented and 'very weak', whereas those of the Kidderminster and Bromsgrove Sandstone formations

Table 7 Strength/consistency/ texture classification of rock, clay and sand.

Rocks	Unconfined compressive strength (UCS) -(MPa)	Description	
	< 1.25	very weak	
	1.25 – 5.0	weak	
	5.0 – 12.5	moderately weak	
	12.5 – 50	moderately strong	
	50 – 100	strong	
	100 – 200	very strong	
	> 200	extremely strong	
Clays	Undrained cohesion (Cu) (kPa)		
	< 20	very soft	exudes between fingers
	20 – 40	soft	moulded by light finger pressure
	40 – 75	firm	moulded by strong finger pressure
	75 –150	stiff	can be indented by thumb
	150 – 300	very stiff	can be indented by thumb nail
	> 300	hard	cannot be indented by thumb nail
Sands	Standard penetration test (SPT) (N value)		
	< 4	very loose	
	4 – 10	loose	can be dug by spade
	10 – 30	medium dense	
	30 – 50	dense	needs pick to excavate
	> 50	very dense	

are generally stronger (Table 6). The sandstones are commonly weakly cemented by calcite, iron oxides and secondary quartz, which breaks down on weathering to leave a 'very weak' rock and ultimately a sandy soil. Typically, variations in strength with depth reflect the variable cementing within alternating beds. The massive sandstones, are commonly separated into large, discrete blocks by open, near-vertical, joints and weak bedding planes. These may be disrupted and further opened by mining subsidence which, together with the local presence of clay infilling in joints, affects the rock mass characteristics of sandstones. Running sand conditions may be encountered where the sandstone bedrock is highly weathered near surface.

The Sherwood Sandstone Group generally provides good foundation conditions and the sulphate content is typically Class 1 (low). The **Kidderminster Formation** is conglomeratic in the lower part; clasts comprise well-

rounded pebbles and cobbles, largely consisting of quartzite. The sandstone matrix is variably cemented, generally weak, although in places calcite cement forms strong rock. The upper part of the formation is predominantly sandstone which is usually rippable, but conglomeratic lenses within the sandstones tend to be stronger. The **Wildmoor Sandstone Formation** has a low clay content, poor cementation, and typically ranges from a 'very weak', speckled red-orange, fine- to medium-grained sandstone to a 'dense' sand, with scattered thin mudstone beds. It is typically diggable. The **Bromsgrove Sandstone Formation** consists mainly of sandstone and pebbly sandstone, which weathers to a silty sand. It is generally a 'weak' rock which is usually rippable, although conglomeratic and well- cemented (calcite) zones within the sandstones tend to be stronger.

Mercia Mudstone Group

In its unweathered state the Mercia Mudstone may be described as an intact, jointed, 'weak' rock, whereas in its fully weathered state it is a reddish brown, 'very soft' to 'hard' silty clay, but commonly containing less-weathered mudrock clasts. The depth of weathering can be considerable, exceeding 30 m in some cases; however, it is more typically 10 to 15 m. The weathering profile is usually progressive, with strength and stiffness tending to increase gradually with depth. However, alternating mudstone and clay, that is, bands of different weathering grade, may be encountered. In most engineering applications the weathering classification (Table 9) of Chandler (1969) is used to classify the Mercia Mudstone prior to

Table 8 Classification of sulphates in soil and groundwater (Anon, 1981a).

Class	Total SO$_3$ (%)	SO$_3$ g/l in 2:1 Soil:Water	SO$_3$ in groundwater g/l
1	< 0.2		< 0.30
2	0.2–0.5		0.3–1.2
3	0.5–1.0	1.9–3.1	1.2–2.50
4	1.0–2.0	3.1–5.6	2.5–5.0
5	> 2	> 5.6	> 5.0

Table 9
Engineering
geological
weathering
classification of
the Mercia
Mudstone
('Keuper Marl')
from Chandler
(1969).

State	Zone	Description	Notes
Fully weathered	IVb	Matrix only	Can be confused with solifluction or drift deposits, but contains no pebbles. Plastic, slightly silty clay. May be fissured
Partially weathered	IVa	Matrix with some claystone pellets, usually about sand size	Little or no trace of original (zone 1) structure, although clays may be fissured
	III	Matrix with numerous lithorelicts, becoming less angular with increasing weathering	Moisture content of matrix greater than that of lithorelicts
	II	Angular blocks of unweathered marl with virtually no matrix	Spheroidal weathering. Matrix starting to encroach along joints; first indications of chemical weathering
Non-weathered	I	Mudstone (generally fissured)	Moisture content varies due to depositional variations

geotechnical testing, sometimes with minor modifications to suit local conditions (CIRIA, 1973).

The weathering zonation is important in the assessment of foundation suitability. Frequently, it is difficult to obtain undisturbed test specimens for strength and deformability tests, and estimates are made from in-situ Standard Penetration Test, cone penetrometer, or more recently self-boring pressuremeter tests. Zone IVb material is typically 'very soft' to 'firm' and has widely varying geotechnical properties, in particular strength and stiffness. In some cases Zone IV material is described as 'friable' and even 'granular'. Zone IVb may be confused with head derived from Mercia Mudstone, or glaciolacustrine clay. The Mercia Mudstone is found to be 'water-softened' where its upper boundary provides a ponding effect, particularly beneath sandstone or made ground (fill). Here it can be expected to have low strength and high deformability. An undrained cohesion range of 12 to 600 kPa is reported for zones II to IV in the Coventry area (Forster, 1991); with considerable overlap across the zones. Little or no Zone I material is recorded in engineering site investigations.

Weathering tends to increase the clay fraction and hence the plasticity, reduce the density, and remove the beneficial mechanical effects of overconsolidation. There is a notable increase in plasticity from zones III to IV. In general, zones I to III have a 'low' to 'intermediate' plasticity, whilst Zone IV has an 'intermediate' to 'high' plasticity (Chandler, 1969). Natural moisture contents are usually close to the plastic limit. Sometimes the Mercia Mudstone, and lithorelicts within it, are described as having a 'shaley' fabric. The siltstones within the Mercia Mudstone are usually well cemented, fine grained, and have a conchoidal fracture. Weathering may occur in cuttings over a period of decades.

Gypsum is encountered as veins and joint infills. In the Cheshire area, leaching out of thick beds of gypsum has resulted in a brecciated fabric (Marsland and Powell, 1990). However, in the district, the primary gypsum content is much lower, and the mineral is generally leached by groundwater dissolution within 30 m of the ground surface.

The Mercia Mudstone tends to have an open, sensitive, microstructure. This suggests that damage to the microstructure can be caused by sample disturbance associated with drilling which has the effect of reducing deformation moduli and, to a lesser extent, strength. Clay-size particles tend to be aggregated into silt-sized peds or clusters. This is suggested in the results of particle-size analyses, where percentage clay values obtained from British Standard (BS1377) methods are smaller than the true value (Dumbleton and West, 1966; Davis, 1967). Davis (1967) quotes true values for percentage clay-size content of 60 to 100%, compared with measured values of 10 to 40%, and quotes values for aggregation ratio, Ar (ratio of clay content from mineralogical analysis to clay content from particle size analysis) of 1.4 to 10.0. These results, if correct, suggest that the measures employed to disaggregate samples in the standard BS1377 particle-size analysis test (Anon, 1990b) may be only partially successful in the case of the Mercia Mudstone. Chandler (1969) suggests that disaggregation is mainly a problem with unweathered Mercia Mudstone, and cites carbonate as one possible cause.

Swelling and shrinkage are two key aspects of the relationship between volume change and water content of clays. However, they are properties rarely determined, at least on *undisturbed* samples, in the course of routine site investigations in the UK. This means that reliance has to be placed on estimates or correlations from other index parameters, such as liquid limit, plasticity index, and density. The Mercia Mudstone is generally considered to have a low swelling potential, due to its low content of recognised swelling clay minerals and pyrite, and also perhaps to its aggregated structure. The presence of gypsum may result in large volume changes, irrespective of clay mineralogy. However, as a result of dissolution and leaching, sulphates tend not to be prevalent in the Mercia Mudstone within the area at outcrop, but occur at depth generally greater than about 30 m. Swelling of compacted samples of Mercia Mudstone was found by Chandler et al. (1968) to be a function of placement moisture content and liquid limit. Difficulties in the compaction of the Mercia Mudstone, and other mudstones are reported (Old et al., 1990).

Excavation may be achieved by digging, and by ripping in the harder horizons. Excavation stability and bearing capacity are generally good, though this depends on local conditions. Fill slopes are usually engineered at 2:1. Sulphate content ranges from Class 1 (low) to Class 5 (very high), but tends to be Class 1 near surface (Forster, 1991). High sulphate contents will attack concrete. Obtaining undisturbed test samples from core is often difficult without the use of triple barrel technology. Increasing use is made of down-hole pressuremeter tests and plate-loading tests, in order to obtain strength and deformation parameters.

Superficial deposits (drift)

TILL AND SANDY TILL

This deposit consists mainly of a silty clay matrix containing exotic clasts of gravel to boulder size, generally less than 10 m thick, but locally up to 40 m thick in buried channels (Millfield and Moxley channels; Tame valley (Figure 28)). The natural moisture content is generally at or below the plastic limit, which indicates the overconsolidated nature of the tills which are generally classified as lodgement tills (Cannell, 1982). A widespread type is an unsorted stony, grey or red-brown sandy clay with rock debris ranging from large boulders to rock flour (Horton, 1975). The deposits are themselves largely unbedded, but may be interbedded with other superficial deposits. Plasticity is in the main 'low' but with some 'intermediate'. Undrained strength classification (based on laboratory tests) varies widely from 'soft' to 'very stiff', and compressibility 'very low' to 'low'. Density varies over a wide range. Bearing capacity is generally good, but may be poor. Slopes and excavations are usually stable but water-bearing silts/sands cause instability. Diggability may be hampered by large boulders. The sulphate content is typically Class 1 (low), and rarely Class 2 (Forster, 1991).

GLACIOFLUVIAL DEPOSITS (SAND AND GRAVEL)

This deposit consists of oxidised, reddish brown, sand and gravel with some paler sand, silt, and clay beds. The deposit is heterogeneous and generally well graded, but individual beds may be poorly graded. The texture ranges from 'loose' to 'dense', with Standard Penetration Test (SPT) values typically in the range 10 to 50. The deposit tends to form sheets and lenticular bodies, the thickness of which is difficult to predict. The sulphate content is typically Class 1 (low), but may be Class 2 (Forster, 1991). Running sand conditions may be encountered.

GLACIOLACUSTRINE DEPOSITS (SAND, SILT, AND CLAY)

These consist mainly of normally consolidated silty clay and clayey silt, with the addition of sand, gravel, and peat, and gravelly, silty, or clayey sand. They tend to have 'low' to 'medium' compressibility although in a few cases 'high', and 'low' to 'intermediate' plasticity; the sands having the more favourable engineering properties. The deposits are generally 'soft' to 'stiff', with a higher moisture content and lower density than the tills. The deposits are suitable as fill or resource (see Minerals Resources section), but provide poor foundation material. Poorly graded silts and sands predominate. Diggability is good. Trafficability is poor when wet. The sulphate content is typically Class 1 (low).

RIVER TERRACE DEPOSITS

These deposits consist mainly of gravel and sandy gravel, but may have a moderate clay component. They are generally 'dense' or 'medium dense' (Powell et al., 1992).

HEAD (SOLIFLUCTION AND GELIFLUCTION DEPOSITS)

Head is a widespread and highly heterogeneous deposit, which may, in some forms, be difficult to distinguish from the weathered source rock. It is usually found at the surface but in many places is buried beneath made ground/fill, and, locally, alluvium. It may consist of reworked in-situ and transported bedrock and superficial material having a wide range of geotechnical properties, though broadly these tend to reflect those of the dominant parent material. Not all deposits of head are necessarily mapped. Head tends to be either sheet-like or to infill hollows and be highly variable in thickness. It may contain relict shear planes where situated on slopes, even of a shallow angle. These may be re-activated naturally or by engineering operations, and represent a hazard. Head may vary from a cohesive, 'soft' to 'stiff', plastic clay to a 'loose' to 'dense', non-cohesive, granular deposit containing gravel or rock fragments, depending on source material. Head is usually unsuitable for use as a foundation material having a poor bearing capacity. Perched water tables and running sands may be present. It may provide a suitable fill material. The sulphate content is typically Class 1 (low).

LACUSTRINE DEPOSITS (INCLUDING PEAT)

These deposits consist predominantly of clay, which may be silty, sandy or gravelly, and, more rarely organic. Engineering problems caused by peat and highly organic silt are due to their low strength, very high compressibility, high (often acidic) water content, and methane generation. The deposits tend to have a higher moisture content and lower density than the tills. Poorly graded silts and sands predominate. Diggability is good. Trafficability is poor when wet. The sulphate content is typically Class 1 (low).

ALLUVIUM

The alluvium encountered in the area is variable in composition because of the variety of both the source rocks within the catchment areas and the alluvial depositional environments. Geotechnical data reflect this variability. It is typically a 'soft' to 'stiff' (more rarely 'hard'), grey or reddish brown, fine-grained sandy or silty clay with impersistent layers or lenses of sand and gravel, and in places contains lenses of brown, amorphous peat (up to

0.5 m thick). Plasticity is generally 'low' to 'intermediate', but may be 'high'. The thickness of the alluvium is variable, as it occupies present and former river channels, but is typically between 2 and 6 m. Engineering behaviour is highly variable, and the alluvium is not always suitable as fill. Excavatibility by digging is good. Stability in excavations is poor. Inflow of water is a problem below the water table. The sulphate content is typically Class 1.

LANDSLIP (AND MASS MOVEMENT DEPOSITS)

Landslip and slope instability are not significant issues in the district. However, excavation of, or construction on, natural slopes, or exceptional climatic conditions may disturb the natural equilibrium, and may present problems in certain circumstances. Heterogeneous deposits such as made ground and head are also susceptible to landslip where they are present on steep slopes, such as the areas adjacent to the Rowley Regis dolerite intrusion. Other formations may exhibit very shallow (less than 1.0 m) instability or hill creep, where weathered or disturbed.

Areas prone to **landslip** are those where slopes are formed on deposits which are dominantly clays or mudrocks, or where there are thin water-bearing sandstones overlying clays and mudrocks, or interbedded with them. Examples of these are Halesowen Formation on Etruria Formation, and Coal Measures. Landslip is also possible on slopes in made ground, particularly colliery waste tips. Such landslips may initiate in the tip but involve slip planes within the underlying natural deposit. Few landslips have been recorded in the area, most being discovered during site investigations or construction work. Landslips have been noted on the Etruria Formation at Bury Hill, Oldbury [972 894], at Brades Village [977 898], at Tansley Hill [961 892], at Haden Cross, Blackheath [963 854; 967 855] and in dolerite at Turner's Hill [965 881; 969 887]. The Bury Hill landslip is formed in periglacially disturbed Etruria Formation (head) to depths of 10 m, on a maximum slope of 10 to 13°. It is complicated by the presence of colliery spoil (made ground) and head, and the possibility of local mining subsidence (Hutchinson et al., 1973). It is possible that similar landslips exist elsewhere in the area of the Rowley Regis dolerite intrusion. The presence of relict slip surfaces in head deposits makes natural landslip, or instability during engineering operations, a distinct possibility even on shallow slopes. Shallow Pleistocene, or relict historical landslips may be difficult to distinguish on the ground as their surface expression is likely to be degraded or obscured by urban development, farming or erosion.

MADE GROUND, WORKED GROUND AND LANDFILL

There is a considerable extent and variety of man-made deposits, particularly in the western part of the district.

Intense industrialisation, followed by decline and dereliction, and the recent proliferation of road construction, has resulted in a complex picture in the urban areas. The geotechnical properties are difficult to predict, and depend on the source material and method of placement; for example, mine waste will have very different properties from domestic waste, and controlled, selective backfilling from uncontrolled tipping. Significant contamination may be present in areas of previous industrial activity, for example former collieries, gas, chemical and sewerage plants.

Moisture content, density, and plasticity tend to have a very wide range. Texture ranges from 'very loose' to 'dense'. The median liquid limit value quoted for undivided made ground and fill by Powell et al. (1992) is 43%. It is notable that little or no positive correlation between standard penetration value and depth appears to exist for made ground and fill (Forster, 1991). Colliery spoil, being the most widespread form, tends to dominate the available data. Colliery waste may be subject to spontaneous combustion and give off noxious fumes. The sulphate content of weathered unburnt colliery waste may be high (Bell, 1981). Alternating granular and cohesive layers may be encountered, and contain localised perched water tables. Fill has been encountered to depths of 20 m.

The use of former quarries and pits for landfill waste disposal may present a hazard from the generation of noxious or explosive gases such as methane, carbon monoxide and carbon dioxide (Dutton et al. 1991; Edwards, 1991), and from toxic leachate in groundwater (Eglington, 1979). Liquid toxic residues may be present either as a primary component of the fill, or generated by secondary chemical or biological reactions within the fill. Fluids and gases can migrate within the fill or into adjacent permeable strata, either within pore spaces of the surrounding deposit or along joints, fractures and faults. Landfill sites situated on the unconfined Sherwood Sandstone Group aquifer or glaciofluvial and alluvial sand and gravel deposits are particularly at risk, especially if the site is in hydraulic continuity with the aquifer (see Water Resources section). Constructional fill (made ground) comprises material that is deposited on the original land surface; it may be spread out to form a general rise in the land surface level. In addition to the factors noted above, former excavations may present constraints to development related to the geotechnical problems of variable fill materials, poor drainage and variable compaction of the fill material (see Foundation Conditions). Many of the sand and gravel quarries along the Tame Valley, near Hams Hall, were backfilled with pulverised fuel ash derived from the former power station at that site; others are flooded or partly flooded.

Guidelines for the control and treatment of landfill sites are set out in Department of the Environment publications (DoE, 1987; 1989), ICRCL (1988) and HMIP (1989). A knowledge of the extent and type of former excavations can resolve many of the problems associated with development; the BGS holds archival information (Harris et al., 1994) on former and current quarrying operations.

UNDERMINING

Shallow and deep colliery-based mining (coal, fireclay and ironstone) in the South Staffordshire and Warwickshire coalfields have left a legacy of shallow workings, shafts, adits and backfilled opencast pits which present problems for land use, such as mining subsidence, shaft collapse and poor foundation conditions (Table 2). Furthermore, much of the early mining in the northern part of the South Staffordshire Coalfield, prior to the 1872 Coal Mines Regulation Act, was unrecorded and uncontrolled; many shafts and mines are recorded inaccurately.

Subsidence is assumed to have taken place soon after mining where longwall methods were used (Healy and Head, 1984) as in the more recent mines east of the Eastern Boundary (Great Barr) Fault, e.g. Sandwell Park and Hamstead collieries in the South Staffordshire Coalfield, and in the eastern part of the Warwickshire Coalfield.

In addition to colliery-based mining, underground mining of Silurian limestone took place at a number of sites in the Black Country (Ove Arup and Partners, 1983; Powell et al., 1992). The constraints placed on planning and development of undermined limestone areas are similar to those outlined above, namely general subsidence due to collapse of voids, and the upward generation of crown holes. An intensive programme involving the investigation, monitoring and treatment of mined limestone cavities is continuing with the support of local authorities and the DoE (Ove Arup and Partners, 1983).

Guidance for engineers and planners involved with the planning and development of undermined sites is given in Healy and Head (1984) and DoE (1988c). Records of shafts and abandoned mines are lodged with the Coal Authority (see Chapter Ten), who should be consulted for detailed information prior to the development of a site in the coalfield. Additional records are held by the BGS, Keyworth, and the technical staff of the appropriate local authority.

The long history of **colliery-based** mining has left a legacy of shallow workings, shafts, adits and backfilled opencast pits which present problems for land use (Figure 3). Hazards include mining subsidence, shaft collapse, flooding, groundwater pollution, gas accumulation, and poor foundation conditions. The wide variety of mining methods including 'bell-pits', 'pillar and stall', 'square works', and 'longwall' working, and variable depths of working produce different engineering problems. The longwall method tends to be more predictable and its extent better documented, compared with the others. Mine gases such as methane, carbon dioxide, and carbon monoxide may accumulate in any

voids remaining after collapse of workings, and migrate through faults, shafts, or permeable strata (Powell et al., 1992). Rising groundwater may pressurise gas and cause it to migrate towards the surface.

Underground mining of **Silurian limestone** is considered in the section on mineral resources. The constraints placed on planning and development of areas underlain by abandoned underground limestone mines (Figure 3) are similar to those outlined above for colliery-based mining. The principal risk is subsidence due to collapse of voids (Ove Arup and Partners, 1983). Crushing of mine pillars, roof collapse and floor heave contribute to the infilling of mined voids through time, but sudden collapse of the roof may cause upward migration of the voids as a chimney-shaped conduit, until the surface is reached when a crown-hole is formed. The latter case has occurred in the past in the Dudley area, but is uncommon in the district due to the greater depth of the mines (Table 3). The Ove Arup and Partners study has shown that about two-thirds of the identified disturbances occurred above mines where the depth of the roof is less than 30 m. General subsidence is largely restricted to mines deeper than 70 m. An intensive programme involving the investigation, monitoring and treatment of the mined cavities has been carried out. The most common form of treatment is to infill the voids by injecting grout comprising gravel, sand, pulverised fuel ash or colliery waste, together with cement in various proportions. The intention is to prevent block falls of the roof and spalling of the pillars, thereby reducing the risk of sudden collapse.

CONSERVATION

Exposures of rocks and their included fossils, and superficial deposits, which demonstrate the geology and geomorphology of the area are an important resource for education, research and recreation. Sites can be protected by their designation as a Site of Special Scientific Interest (SSSI) or as a Regionally Important Geological/ Geomorphological Site (RIGS), both administered by English Nature (Chapter Ten). Planners and developers need to be aware that, given suitable treatment such as landscaping, clearing and stabilising rock faces, the presence of a rock section may enhance the amenity value of a site. Landfill operations can totally obscure unique or important geological features; where backfilling is deemed necessary, the most valuable features may be preserved by leaving part of the quarry face clear.

At the time of writing there is one geological site of Special Scientific Interest (SSSI) in the district: Bromsgrove Road Cutting, Tenterfields [SO 971 835]; exposures of Halesowen Formation sandstone.

THREE

Cambrian and Ordovician

The only Cambro-Ordovician formation to crop out in the district is the Lickey Quartzite, the precise age of which is uncertain; it was formerly correlated with Cambrian strata but is now considered probably to be Ordovician in age. Undoubted Cambrian and Ordovician strata are, however, proved at depth in the eastern part of the district (Figure 6) and are inferred, largely on seismic evidence, to underlie the Knowle Basin (Chapter Nine). These concealed Cambro-Ordovician rocks (Table 10) are a westward extension of the succession exposed in the Nuneaton area and underlying the Coal Measures of the Warwickshire Coalfield (Bridge et al., 1998).

HARTSHILL SANDSTONE FORMATION

The Hartshill Sandstone Formation, present in the Nuneaton area of the Coventry district (Sheet 169) (BGS, 1994), consists mainly of hard, quartzose sandstone, more than 250 m thick. Partial stratotype sections are available at two SSSIs, the basal strata at Boon's Quarry [SP 3279 9471] and the topmost strata at Woodlands Quarry [SP 3248 9480] (Bridge et al., 1998). It represents shallow-marine, tidally influenced coastal deposition, and, towards the top, includes a condensed calcareous and phosphatic unit, the Home Farm Member (Brasier, 1992). Although unproved in the district, fossiliferous clasts from the Permian Clent Breccia in the Nechells Borehole [097 895] (Chapter Ten) indicate derivation from correlatives of the Home Farm Member from a former source area within, or adjacent to, the district. The presence of a seismic reflector attributed to the Hartshill Formation is taken to indicate wide distribution of the formation under the Knowle Basin (Chapter Nine), as shown on Section 1 of the Birmingham Sheet 168 (BGS, 1996).

The Clent Breccia in the Nechells Borehole contains abundant quartzite clasts (Boulton, 1924), together with limestone and calcareous sandstone clasts containing Lower Cambrian fossils that originated from strata of at least two distinct ages (Plate 3).

1. Clasts of crystalline limestone yielded the brachiopods *Micromitra* sp. and *Paterina* sp. 'A' (of Brasier, 1984) and fossils of uncertain affinity: *Camenella* cf. *baltica*, *Coleoloides typicalis*, *Hyolithellus* cf. *micans*, *Torellella* cf. *biconvexa* and *T. lentiformis* (of Missarzhevsky). Blocks of muddy or silty limestone contain *C. typicalis*, *Paterina* sp. 'A', and fragments of Hyolitha tentatively referred to *Allatheca degeeri*, *Burithes alatus*, *Doliutus?*, *Gracilitheca aequilateralis* and *Tuojdachithes?* *biconvexus*. All these taxa are known from the Home Farm Member in the upper part of the Hartshill Sandstone Formation (Brasier, 1986), and it is

likely that the clasts were eroded from a similar contemporaneous deposit of the *Camenella baltica* Biozone in or adjacent to the district (Brasier, 1992).

2. A younger deposit was the source of a single clast (at least 80 mm across) of muddy limestone that contains the brachiopod *Obolus parvulus* and the trilobites *Runcinodiscus* sp. and *Strenuella?* sp. juv. This is correlated with the middle units of the Comley Limestones (*Strenuella* or *Protolenus* Biozone) of Shropshire (Cobbold, 1921), and is presumably contemporaneous with the middle part of the Purley Shale Formation which overlies the Hartshill Formation in the Nuneaton succession (see also below). It is unlikely that this clast was transported from a site as distant as Shropshire, and it is thought that a more local Cambrian succession included calcareous strata at the level of the *Strenuella* or *Protolenus* Biozone, even though no similar calcareous beds are known at the same level in the Nuneaton area.

The quartzite proved at the base of the Walsall Borehole is considered below.

STOCKINGFORD SHALE GROUP

The Stockingford Shale Group is present at depth in the eastern part of the district, to the east of the western boundary fault of the Warwickshire Coalfield. At Nuneaton in the Coventry district (Bridge et al., 1998) the Stockingford Shale Group is more than 800 m thick and ranges in age from early Cambrian (Comley) to early Ordovician (Tremadoc). The Cambrian part of the succession represents moderately shallow-water deposition on a stable shelf, but during the Tremadoc the development of half-grabens allowed the rapid accumulation of thick deposits of deeper water mud (Smith and Rushton, 1993).

The succession proved in boreholes corresponds to the upper part of the group at outcrop in the Nuneaton area, namely the upper four formations (Table 11).

The **Outwoods Shale Formation**, of early Merioneth age, consists of alternations of dark grey laminated mudstone, pyritic and carbonaceous, with grey bioturbated mudstone and minor sandstone. Reference sections are exposed in Purley Quarry [SP 305 961] and Oldbury Quarry [SP 310 950]. The formation is described fully by Taylor and Rushton (1971, pp.9–22) and Bridge et al. (1998). Although the formation has not been proved in the district, it was encountered in the Trickley Lodge Borehole [SP 1603 9884] located about 2 km to the north (Figure 6); the occurrence of *Homagnostus obesus* proves the presence of the *Olenus* Biozone. It is possible that the Outwoods Shale Formation is present at depth

Figure 6 Sketch map showing sub-Carboniferous crop of Cambrian and Ordovician formations at the eastern end of the district, as inferred from borehole information.

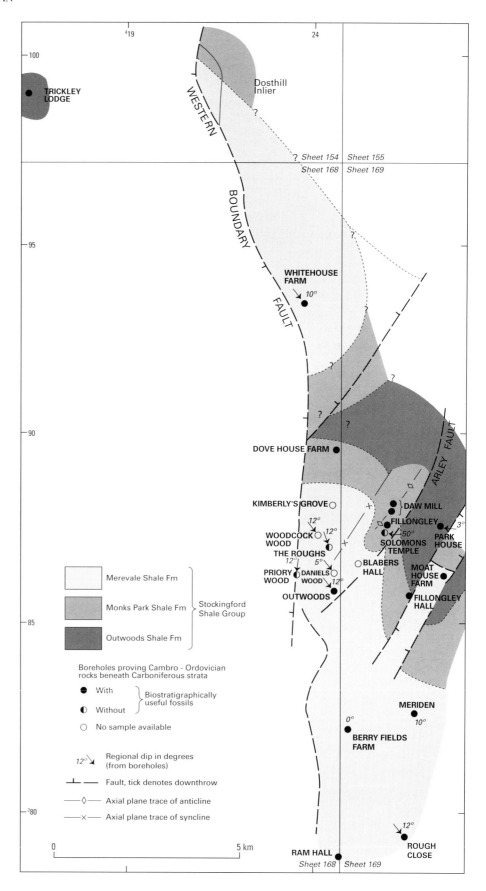

Table 10
Stratigraphy of
the Cambrian
and Ordovician
strata in the
district and
adjacent area.

Chrono-/Biostratigraphy		Lithostratigraphy	
System/Series	Zones	Formation	Group
ORDOVICIAN — Arenig ? or younger		Lickey Quartzite*	
ORDOVICIAN — Tremadoc		Barnt Green Volcanic Formation*	
		~~ ? ? ? ~~~~	
	Rhabdinopora flabelliformis	Merevale Shale Formation†	Stockingford Shale Group
	'basal Tremadoc'		
CAMBRIAN — Merioneth	*Acerocare*	Monks Park Shale Formation†	
	Peltura scarabaeoides *Peltura minor* *Protopeltura praecursor*		
	Leptoplastus		
	Parabolina spinulosa	Moor Wood Sandstone Formation	
	Olenus	Outwoods Shale Formation	
	Agnostus pisiformis		
CAMBRIAN — St David's	*forchhammeri*	Mancetter Shale Formation	
	paradoxissimus	Abbey Shale Formation	
	oelandicus		
CAMBRIAN — Comley	*Protolenus*	Purley Shale Formation	
	Strenuella		
	Callavia		
	Camenella baltica	Home Farm Member / Hartshill Sandstone Formation (part)	
	Aldanella attleborensis		

* indicates formations proved in the South Staffordshire area of the district.

† indicates formations proved in the Warwickshire area of the district.

~~ unconformity.

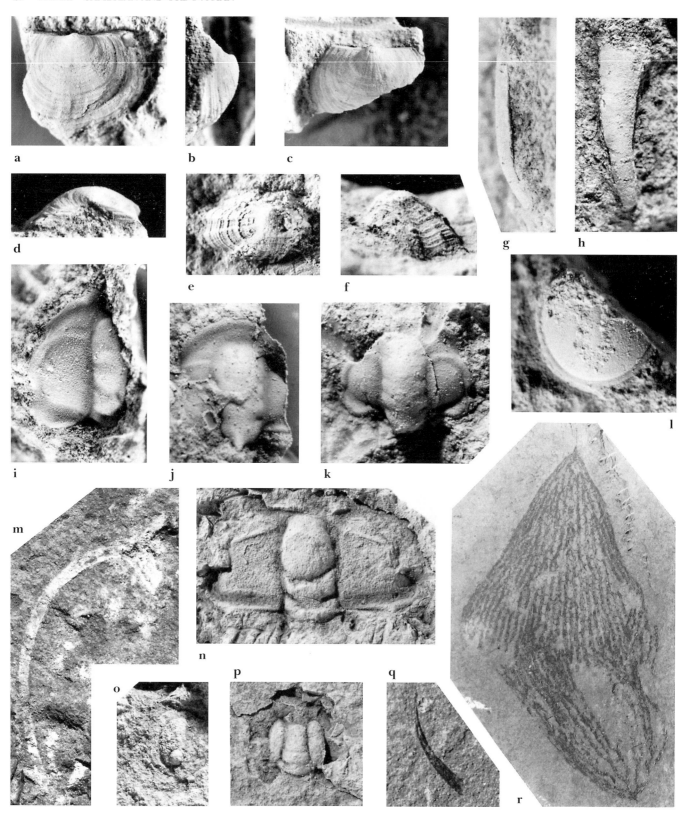

Plate 3   Examples of Cambrian and Ordovician fossils from the district. Specimen numbers prefixed BU are from the Lapworth Museum, University of Birmingham. All the rest are in the collections of the British Geological Survey.

Table 11 Cambrian and Ordovician strata proved in boreholes in the Warwickshire coalfield

Formation	Thickness in metres	Series
Merevale Shale	about 100 seen at outcrop; concealed thickness much greater	Tremadoc (Ordovician)
Monks Park Shale	80	
Moor Wood Sandstone	15, thinning out westwards	Merioneth (Upper Cambrian)
Outwoods Shale	250 (excluding sills)	

between the Dove House Farm and Whitehouse Farm boreholes (Figure 6) because the former proved a level towards the base of the overlying Monks Park Shale Formation.

Plate 3 *continued*

Figures a–l are from clasts in the Clent Breccia in the borehole at Nechells Gasworks [097 895]; Figures a–h are from the *Camenella baltica* Biozone (approximately upper Tommotian to lower Atdabanian); Figures i–l are from the *Strenuella* Biozone (approximately upper Atdabanian or lower Botomian).

a–d, *Paterina* sp. 'A' [of Brasier 1984], from clasts at 498 feet (151.7 m) depth. a, b, d, ventral, lateral and posterior views of a pedicle valve, AT 1126b, × 5. c, internal mould of pedicle valve, AT 1126c, × 5.

e, f, *Camenella* cf. *baltica* (Bengtson), top and oblique views of mitral sclerite, AT 1123, × 10. Clast at 465 feet (141.7 m).

g, h, *Torellella lentiformis* (Syssoiev) [of Missarzhevsky], side view showing sharp lateral edge and proximal curvature, and ventral (?) view, AT 1121, × 10. Clast at 465 feet (141.7 m).

i–k, *Strenuella?* sp. juv. i, latex cast of BU 2493. j, k, latex cast of BU 2494 and internal mould of counterpart, BU 2492. All × 10, all from one clast at 520 feet (158.5 m) depth.

l, *Runcinodiscus* sp., small pygidium, BU 2496, × 10. From clast at 520 feet depth.

m–p, all collected by Dr F Raw from the Monks Park Shale Formation, Dosthill Quarry [approximately SP 211 998]. *Peltura scarabaeoides* Biozone, *bisulcata* Subzone.

m, n, *Ctenopyge* (*Ctenopyge*) *bisulcata* (Phillips), librigena Zs 6301 showing straight posterior border and long curved spine, and cranidium Zs 6296, both × 6.

o, p, *Sphaerophthalmus humilis* (Phillips), semi-circular librigena Zs 6304 (genal spine missing but showing globular eye lying far back), and cranidium Zs 6308, both × 6.

q, *Phakelodus* sp. single element, Pl 3766, × 8. Base of Merevale Shale Formation, Dosthill Quarry [approximately SP 211 998]. Zone uncertain, but near the Merioneth-Tremadoc boundary, above the *scarabaeoides* Biozone and below the *flabelliformis* Biozone.

r, *Rhabdinopora flabelliformis* cf. *socialis* (Salter) [of Bulman and Rushton 1973], BKE 8289, × 2. Whitehouse Farm Borehole (491.8 m), Merevale Shale Formation, lower Tremadoc, *flabelliformis* Biozone.

The **Moor Wood Sandstone Formation**, though present in the Nuneaton area, appears to thin out to the west (BGS, 1994), and is not proved in the district.

The **Monks Park Shale Formation** consists of dark grey mudstones in the lower part and soft black carbonaceous and pyritic mudstone above, and represents condensed deposition in conditions of low oxygenation (Taylor and Rushton, 1971 pp.23–34; Bridge et al., 1998). The core of the Merevale No 1 Borehole provides the best reference section. The Dove House Farm Borehole [2474 8912] encountered a level near the base of the formation, at which the presence of *Orusia lenticularis* indicates the *Parabolina spinulosa* Biozone. Just east of the district higher parts of the formation were encountered along the Fillongley Anticline, notably in boreholes at Daw Mill: Borehole 200RI [2632 8724] yielded *Parabolina spinulosa*, indicating the *spinulosa* Biozone, and Borehole 122 [2633 8760] yielded several fossils representing parts of the *Peltura minor* and *P. scarabaeoides* biozones and perhaps higher horizons also (Bridge et al., 1998). The *scarabaeoides* Biozone is also proved at outcrop in the Dosthill Inlier [SP 212 995], immediately north of the district (Figure 6; Plate 3, m–p).

The **Merevale Shale Formation** is the most widely distributed at depth and is inferred, in its fullest development, to be much thicker than the 90 m thickness cropping out in the Merevale area (Taylor and Rushton, 1971, p.34), where only the basal part of the formation is preserved. The basal stratotype is taken in the core of the Merevale No. 2 Borehole. It consists of grey and greenish grey mudstone, commonly bioturbated, with minor siltstone and sandstone beds (Bridge et al., 1998). Such beds are proved at depth beneath the Coventry and Warwick districts. The faunal succession ranges up through the lower half of the Tremadoc Series (Old et al., 1987, p.3) and, at the base, includes beds older than any known in the Shineton Shales of Shropshire (Stubblefield and Bulman, 1927).

Several boreholes near the eastern edge of the Warwickshire Coalfield proved Merevale Shale Formation, either the 'Basal Tremadoc' of Old et al. (1987, p.3) or the *Rhabdinopora* [*Dictyonema*] *flabelliformis* Biozone. The faunas are detailed in Bridge et al. (1998). There appears to be a general southward younging from the Dove House Farm Borehole, across the Fillongley Anticline and Arley Fault towards the *flabelliformis* Biozone identified in the Ram Hall [2469 7809] and Rough Close [2648 7851] boreholes (Figure 6), and thence towards the younger Tremadoc proved in the Warwick district. However, there are complications, such as the *flabelliformis* Biozone in the Outwoods Borehole [2462 8528] (Figure 6) and the 'Basal Tremadoc' in the Berry Fields Farm Borehole [2499 8125], that indicate that the structure is not so simple. North of the Dove House Farm Borehole, records are sparse. The *flabelliformis* Biozone is present in the Whitehouse Farm Borehole [2387 9303] (Plate 3, r) and Merevale Shales containing the brachiopod *Eurytreta sabrinae*, and presumably referable to the 'Basal Tremadoc', are mapped in the Dosthill inlier, immediately to the north (Taylor and Rushton, 1971, pp.35, 54).

LICKEY QUARTZITE

The Lickey Quartzite crops out in an inlier that extends from Kendal End in the Redditch district [SP 001 746] north-north-westwards to around Holly Hill [SO 991 784]; only a small part of the outcrop lies in the district. The formation consists predominantly of hard, quartzose sandstone, commonly much shattered. A large quarry in Rednal Gorge [998 759] (Old et al., 1991) provides a reference section. The base of the formation is not seen, and folding and faulting are such that the exposed thickness is unknown. Although the structure also is not well known, it appears broadly to be an anticline with an axial plane trace that trends north-north-west. At outcrop near Rubery Station [993 772] the Lickey Quartzite is overlain unconformably by the Llandovery Rubery Sandstone and the Carboniferous Halesowen Formation.

The Lickey Quartzite was described fully by Old et al. (1991). Petrographically it varies from lithic arenite to quartz arenite (Strong, in Old et al., 1991, p.5), occurring in massive beds locally, separated by thin partings of sandy mudstone.

The Lickey Quartzite was formerly assigned to the Lower Cambrian (Comley Series) by lithological comparison with the Hartshill Sandstone (or 'Quartzite') formation of the Nuneaton area, the age of which is well established (Brasier, 1986). However, the recovery of microfossils from a mudstone bed in the Lickey Quartzite enabled Molyneux (in Old et al., 1991, p.4) to propose a post-Cambrian, probably Ordovician, age. He also showed that the presumably underlying Barnt Green Volcanic Formation was not Precambrian (Neoproterozoic), as formerly supposed, but possibly Tremadoc in age. In view of the revised age of the formation it may be more appropriately correlated with the Stiperstones Quartzite Formation (of Arenig age) in the Shelve area of Shropshire and Powys (Whittard, 1979), rather than the Hartshill Sandstone Formation.

QUARTZITE OF THE WALSALL BOREHOLE

The Walsall Borehole [SP 009 980], just north of the district, proved 10 m of unfossiliferous pale quartzite (base not seen) underlying Llandovery (Telychian) shale (Table 12). Butler (1937, p.249) inferred a Cambrian age, comparing it to quartzites of 'the Lickeys and other Midland outcrops', and correlation with the Lickey Quartzite has subsequently been accepted (Cocks et al., 1992). Although such a correlation would now, following the likely assignment of the Lickey Quartzite to the Ordovician, imply a post-Cambrian age for the Walsall Borehole quartzite, the possibility of an early Cambrian age is maintained by the occurrence in the Clent Breccia in the Nechells Borehole (see above) of large clasts of pale quartzite associated with fossiliferous Lower Cambrian limestone clasts (Boulton, 1924). In the absence of faunal or other evidence the Walsall Borehole quartzite may be either Cambrian or Ordovician in age.

FOUR

Silurian and Lower Devonian

In the west of the district, Silurian rocks crop out locally at Great Barr and Rubery and have been proved to occur extensively below Coal Measures in the South Staffordshire Coalfield. Within the Knowle Basin, of the central part of the district, the nature of the strata beneath the thick Permo-Triassic sequence is as yet unknown (see Chapter Nine for discussion). In the Warwickshire Coalfield, in the east of the district, Silurian and Devonian rocks are absent so that Carboniferous strata rest unconformably upon Upper Cambrian and Lower Ordovician rocks.

In the South Staffordshire Coalfield, Silurian rocks rest unconformably upon quartzite, generally assigned to the Lickey Quartzite, of probable Ordovician age, but alternatively of possible Cambrian age (see Chapter Three). A conformable sequence ranging from late Llandovery to Přídolí or possibly Early Devonian age is present in the coalfield (Whitehead and Eastwood, 1927), although only parts of this sequence have been proved within the district. The position of the Silurian–Devonian boundary in Great Britain is at present uncertain, but possibly occurs within the Ledbury Formation (White and Lawson, 1989), which is the youngest pre-Carboniferous formation present in the district.

Following uplift of the area during the Ordovician, the district formed part of a landmass which lay at the south-eastern margin of the Iapetus Ocean. A marine transgression during Llandovery times resulted in the eastward migration of the tropical or subtropical sea across this landmass, with deposition commencing in the district during Telychian times (Ziegler et al., 1968). The earliest Silurian deposits in the district, the Rubery Formation, contain brachiopod communities which indicate deposition in moderate depths of sea water (Ziegler et al., 1968). Deposition during Wenlock and much of Ludlow times was dominated by shallow water mudstones and siltstones of the Buildwas, Coalbrookdale, Lower, Middle and Upper Elton (Lower Ludlow Shale) and Whitcliffe (Upper Ludlow Shale) formations. These formations are separated by limestone-dominated units, comprising the Barr Limestone, Much Wenlock Limestone and Aymestry (Sedgley) Limestone formations. The limestones mark periods of diminished mud and silt supply to the shelf during slight falls in relative sea-level. A marine regression during late Ludlow to early Devonian times is marked by a gradual change from a marine to continental environment during deposition of the Downton Group (Downton Castle Sandstone, Temeside Shales and Ledbury formations). The last is interpreted as being deposited across tidal flats or fluvial floodplains with marine tidal incursions (White and Lawson, 1989).

RUBERY FORMATION

The Rubery Formation, formerly referred to as the Upper Llandovery Sandstone (Eastwood et al., 1925, p.12), is exposed in a small inlier at Rubery (Old et al., 1991), in the south-west of the district, and in a fault-bounded wedge along the Eastern Boundary Fault at Great Barr, in the north-west of the district. The presence of *Costistricklandia lirata 'alpha'* and '*C. lirata typica*' (now assigned to *Stricklandia laevis* and *Pentameroides*, respectively) at these localities suggests a late Telychian age for the formation (Ziegler et al., 1968, p.764).

At Rubery, the formation rests unconformably upon the Lickey Quartzite, and is, in turn, overlain unconformably and overstepped by the Upper Carboniferous Halesowen Formation. The Rubery Formation comprises a massive, pale grey stained red or purple, coarse-grained sandstone, interbedded with purple, fissile mudstone (the Rubery Sandstone), overlain by buff, grey, blue and purple, non-calcareous, fissile mudstone, interbedded with fine-grained sandstone (the Rubery Shale). This sequence was formerly exposed during widening of the Bristol Road (Wills et al., 1925), located about 400 m to the south of the district, in the Redditch district (Old et al., 1991, p.7). In the Birmingham district, both hard, red sandstones and purple, yellow and grey banded, weathered, fissile mudstones were recorded in boreholes at Rubery Hill Hospital [995 780]. The Rubery Sandstone was formerly exposed at Hollyhill Quarry [9910 7838], resting unconformably upon the Lickey Quartzite [Eastwood et al., 1925, p.13]. The Rubery Shale was formerly exposed at Rubery Station [991 782], overlain unconformably by the Halesowen Formation (Eastwood et al., 1925).

At Great Barr the Rubery Sandstone crops out in a 530 m-long fault-bounded wedge. The formation is poorly exposed at Aston University Sports Centre [0385 9574]. It comprises pale yellow, fine-grained, laminated and thinly-bedded sandstone, dipping 40° to the south-east. Fossils collected from near to this locality during the primary survey, recorded by Jukes (1859, p.110), have not been traced. However, Salter's manuscript list shows that he identified 12 (not 2) specimens of 'Pentamerus' (=*Stricklandia*) *lens*, and one fragment of *Costistricklandia lirata*.

About 117 m of strata attributed to the Rubery Formation were proved in the Walsall Borehole [SP 009 980], to the north of the district (Table 12; Butler, 1937). The formation comprises, at the base, about 39 m of purple and greyish green mudstone with thin beds of shelly limestone, grey siltstone and bentonitic clays, overlain by a 5 m-thick brownish purple, fine-grained, calcareous sandstone; the occurrence of *C. lirata 'alpha'* indicates the *griestoniensis* Biozone. This in turn is overlain by about 73 m of purple,

Table 12 Lower Palaeozoic strata proved in the Walsall Borehole (based on Butler, 1937).

System or Series	Formation unit		Depth (m)	Divisions of Butler (1937)
Wenlock	Much Wenlock Limestone		12.8–24.1	D & E
	Coalbrookdale		24.1–232.9	F to K
	Barr Limestone		232.9–242.5	L
	Buildwas		242.5–265.2	M
Llandovery	Rubery	'shale'	265.2–338.3	N & O
		sandstone	338.3–343.2	P
		'shale'	343.2–381.9	Q to S
Ordovician or Cambrian?	(Lickey or Hartshill?)	quartzite	381.9–391.7	T

maroon, and grey-green mudstone with some thin beds of siltstone and limestone. The boundary with the overlying Buildwas Formation (see below) is gradational.

BARR LIMESTONE FORMATION

The Barr Limestone Formation forms a small outcrop to the west the Eastern Boundary Fault at Great Barr, in the north-west of the district, extending northward onto the Lichfield (154) district. The formation has been equated with, and occasionally referred to as the Woolhope Limestone (Jukes, 1859, p.112; Barrow et al., 1919, p.9; Eastwood et al., 1925, p.12). However, Bassett (1974) proposed that the Barr Limestone correlates with all, or part, of the *riccartonensis* Zone of the Sheinwoodian Stage, and is thus younger than the Woolhope Limestone of the Welsh Borderlands.

At the time of survey, the formation was not exposed in the district, although small quarries in the limestone occur a short distance to the north, in the Lichfield district, from Daisy Bank [SP 0410 9765] to Cuckoo's Nook [SP 0530 9900]. The formation comprises grey, fine-grained limestone, interbedded with grey mudstone with calcareous nodules, and thin bentonitic clays. Fossils collected in the Lichfield district are listed by Jukes (1859, p.112), Barrow et al. (1919, p.12) and Bassett (1974). The base of the formation is marked by a gradational increase in the thickness and number of limestone beds from the underlying Buildwas Formation, whereas the top is marked by a gradational decrease in the thickness and number of limestone beds to the overlying mudstone-dominated Coalbrookdale Formation.

The formation is proved in the Walsall Borehole (Table 12), in the Lichfield district, to be 9.6 m thick (Butler, 1937). The limestone is also believed to have been proved, at depth, in the Birmingham district; in underground boreholes at Heath Colliery, West Bromwich [0073 9119 and 0084 9121], which yielded *Bumastus barriensis* Murchison (King, 1921), and at a depth of 135 m below OD in a gate road from Springfield Colliery No.1 Shaft [SO 958 883], Rowley Regis.

COALBROOKDALE FORMATION

The Coalbrookdale Formation, formerly referred to as the Wenlock Shales (Eastwood et al., 1925, p.11), crops out in the vicinity of Great Barr, to the west of the Eastern Boundary (Great Barr) Fault. Correlation with the Coalbrookdale Formation of the Welsh Borderlands is made by Cocks et al. (1992), who consider the formation to have been deposited during the Sheinwoodian and Homerian stages of the Wenlock Series. The formation is formally defined by Bassett et al. (1975).

The Coalbrookdale Formation comprises greyish green to greenish grey, calcareous mudstone with thin beds of limestone nodules. The formation is fossiliferous; a list of fossils, dominated by benthic, shelly fauna, collected from the adjacent Lichfield district is provided by Barrow et al. (1919, p.13).

The formation is typified by low featureless ground lacking exposures, present to the east of the Russell's Hall and Eastern Boundary faults, where it is unconformably overlain by Coal Measures (King, 1921; Eastwood et al., 1925). However, it is proved in numerous boreholes including the Walsall Borehole (Table 12), in the Lichfield district, where it attains its maximum recorded thickness of 208.8 m (Butler, 1937).

CONCEALED SILURIAN AND LOWER DEVONIAN STRATA

The Buildwas, Aymestry (Sedgley) Limestone, Whitcliffe, Downton Castle Sandstone and Temeside Shale formations (Table 1) do not crop out, and are not proved at depth, within the district. However, from comparisons with the successions present in the neighbouring districts of Dudley and Bridgnorth (167) and Lichfield (154), they are inferred to be present, at depth, in the district.

The **Buildwas Formation**, formerly referred to as the Lower Wenlock Shale, is proved in the Walsall Borehole (Table 12), in the Lichfield district (Butler, 1937). The section comprises 23 m of grey mudstone with thin beds of limestone and limestone nodules (Butler, 1937), with fauna indicating a correlation with the *murchisoni* Zone of the Wenlock Series (Bassett, 1974). This formation has been incorrectly referred to as Rubery Shales by Cocks et al. (1992).

The **Aymestry (Sedgley) Limestone** crops out at Turner's Hill, in the Dudley district (Sheet 167) (Whitehead and Pocock, 1947; Ball, 1951). The limestone, which is about 8 m thick, is correlated by the above authors with the Aymestry Limestone of the Welsh Borderlands. However, the strata at Sedgley are considered to be Gorstian to Ludfordian age (Cocks et al., 1992) and are thus slightly younger than the Aymestry Limestone of the type area.

The **Whitcliffe Formation**, proved at crop in the Dudley district, comprises from 9–15 m of silty shales with limestone nodules and thin sandstone beds (King, 1917; Whitehead and Pocock, 1947; Ball, 1951). The succession, here, is considered to be of Ludfordian age and is correlated with the type area of the Welsh Borderlands (Cocks et al., 1971; 1992).

The **Downton Castle Sandstone**, about 20 m thick, and **Temeside Shales**, about 10 m thick, occur at crop in the Dudley district (King, 1917; Whitehead and Pocock, 1947; Ball, 1951). Both formations are considered to be of Přídolí age (Cocks et al., 1992).

The Much Wenlock Limestone, Lower, Middle and Upper Elton formations and Ledbury Formation do not crop out in the district, but are proved at depth in borehole and shaft records. These formations are discussed separately below.

MUCH WENLOCK LIMESTONE FORMATION

The Much Wenlock Limestone Formation (Bassset, 1989) was formerly referred to as the Wenlock Limestone (Eastwood et al., 1925) and Dudley Limestone (Whitehead and Eastwood, 1927). The formation crops out to the west at Dudley (Sheet 167) (Butler, 1939; Whitehead and Pocock, 1947), and at Walsall, in the Lichfield (154) district (Barrow et al., 1919). The formation forms the top of the Wenlock Series in the West Midlands, although the correlation with the equivalent strata in the Welsh Borderlands remains contentious (Cocks et al., 1992).

The Much Wenlock Limestone (Basset, 1989) is subdivided into three members; the Upper and Lower Quarried Limestone, 9 to 10 m and 12 to 16 m thick, respectively, separated by the Nodular Limestone, which is 31 to 38 m thick. The Upper and Lower Quarried Limestone members consist of hard, bioclastic limestone, locally with large bioherms (patch reefs) consisting of fossiliferous micritic limestone. The Nodular Limestone Member comprises nodular, shelly limestone interbedded with thin partings of grey-green calcareous mudstone and siltstone.

The formation is proved at depth, to the east of the Russell's Hall Fault, in shaft records for Pennant Hill and Knowle collieries (Eastwood et al., 1925) and Yew Tree and Titford Bridge collieries (King, 1921), and in numerous shafts and boreholes in the Willenhall and Darlaston district (SO99NE) and Dudley and Wednesbury district (SO99SE). The formation was worked, at depth, in the district at Blackham, Cow Pasture, Dudley Port, Grovelands, Tividale and other mines (Ove Arup and Partners, 1983; Table 3).

LOWER, MIDDLE AND UPPER ELTON FORMATIONS, UNDIFFERENTIATED (LOWER LUDLOW SHALES)

The formations are the lowest units of the Ludlow Series in the West Midlands. In the Welsh Borderlands, the Lower Ludlow Shales, as they were formerly termed, have been redefined as the Lower, Middle and Upper Elton formations (Lawson and White, 1989). However, the subtle lithological differences between the three formations could not be recognised in South Staffordshire. Furthermore, the use of the epithets, together with the same geographical name is contrary to the North American Stratigraphic Code (NACSN, 1983). Therefore, the formations are referred to as 'undifferentiated' until the status, and correlation with the Ludlow sequence in the Welsh Borderlands is determined.

The undifferentiated unit is proved at crop in the vicinity of Dudley (Whitehead and Pocock, 1947) and Walsall (Barrow et al., 1919). It comprises about 152 m of greenish grey fissile mudstone and sandy mudstone with thin beds of limestone nodules. In the district, the formation is proved, at depth, to the east of the Russell's Hall Fault at Knowle Colliery [9609 8830] (King, 1921) and in boreholes in the Darlaston, Wednesbury and Sandwell areas (sheets SO99NE and SO99SE).

LEDBURY FORMATION

The Ledbury Formation, formerly known as the Ledbury Shales (Eastwood et al., 1925), Passage Beds (Whitehead and Eastwood, 1927) and Red Downton Beds (Cocks et al., 1971), is the stratigraphically highest unit present below the unconformable Coal Measures. The formation is considered to be Přídolí to Early Devonian in age in the type area of the Welsh Borderlands (White and Lawson, 1989).

The formation comes to crop in small inliers at Wollescote and Netherton, in the vicinity of Dudley (Sheet 167), where it comprises about 80 m of red, purple and green, micaceous siltstone and sandstone. In the district, the formation occurs at subcrop below the unconformable Coal Measures to the west of the Russell's Hall Fault. The formation, proved in shaft records for Manor Pit [9766 8311], consists of up to 25 m of red shale and marl with a fauna typical of the Ledbury Formation (King, 1921). Mine workings at Coombes Wood Pit [9715 8461], north-east of Halesowen town centre, prove red shales and mudstones assigned to the Ledbury Formation (compare with King, 1921).

FIVE

Carboniferous and Lower Permian

Carboniferous and Lower Permian strata (Table 13) crop out in the western and eastern parts of the district (Figure 1). The western part lies on the margins of the South Staffordshire Coalfield where approximately 150 m of Coal Measures and 450 m of barren measures (predominantly red) rest unconformably on basement rocks of Silurian to Devonian age. Barren measures onlap and overstep southwards to rest on Ordovician rocks in the south-west of the district (Figures 7 and 8). The eastern part of the district forms the western edge of the Warwickshire Coalfield; approximately 100 m of Coal Measures and 875 m of barren measures rest unconformably upon Cambrian to Ordovician strata. Locally, to the north and east of the district, the Namurian Millstone Grit is present at the base of the Carboniferous succession (Taylor and Rushton, 1971; Bridge et al., 1998).

The boundary between the Lower Coal Measures and the Middle Coal Measures (Table 13) is taken at the base of the Vanderbeckei Marine Band, which also marks the boundary between the Langsettian (Westphalian A) and Duckmantian (Westphalian B) stages (Ramsbottom et al., 1978). The boundary between the Duckmantian and Bolsovian (Westphalian C) stages is taken, regionally, at the level of the Aegiranum Marine Band; the latter is poorly developed in the district, probably as a result of the diachronous passage of the Middle Coal Measures into red-bed lithofacies (Etruria Formation) at about this stratigraphical level.

The Carboniferous succession was deposited during the closing phases of the Variscan orogeny. This coincided with a gradual change in climate from humid to semi-arid conditions by Early Permian times which is reflected in a change from grey-bed, coal-forming deposition to red-bed sedimentation (Besly, 1987; Glover and Powell, 1996). Orogenic hinterland activity to the south of the district also resulted in intermittent sedimentation and localised uplift, folding and erosion (Waters et al., 1994).

COAL MEASURES

South Staffordshire Coalfield

The Coal Measures of the South Staffordshire Coalfield rest unconformably upon strata of Silurian to Early Devonian age (Figures 7 and 8). To the south-west of the Russell's Hall Fault, much of the area of the South Staffordshire Coalfield included in the Birmingham district is underlain by the Ledbury Formation (Přídolí to Early Devonian in age; see Chapter Four). However, to the north-east of the Russell's Hall Fault, the Coal Measures rest unconformably upon strata of Wenlock to Ludlow age and are locally faulted against quartzite of probable Ordovician age (Chapter Three).

Within the district, much of the South Staffordshire Coalfield is concealed by overlying Upper Carboniferous strata to the west of the Great Barr Fault, and by Lower Permian and Triassic strata to the east of the fault. The main outcrops of the Coal Measures occur in the north-west of the district. To the north of the Ball's Hill Fault (Figure 1), both Lower and Middle Coal Measures strike broadly northwards with a consistent westwards-younging pattern interrupted only by gentle north-trending open folds and a series of east- to west-trending faults which parallel the Ball's Hill Fault. A second outcrop of Coal Measures is present in the hanging wall (south) of the Ball's Hill Fault; this forms a broad north-trending and plunging anticline. The south and east of this outcrop is terminated by a series of fault splays of the Great Barr Fault; to the west Coal Measures pass conformably upwards into the Etruria Formation. Minor outcrops of Middle Coal Measures occur to the south of these areas — most of which are, in part fault-bounded; the largest of these is the Coombeswood Inlier which occurs as a small fault-bounded sliver adjacent to the Russell's Hall Fault [975 855] (Figure 1). West of the Great Barr Fault, the South Staffordshire Coalfield is bisected by the Russell's Hall Fault which forms a prominent north-north-west-trending ridge.

The stratigraphy and depositional environments of the Coal Measures in the South Staffordshire Coalfield are described in detail in Eastwood et al. (1925) and Whitehead and Eastwood (1927), and only a brief review is presented here (Figures 7 and 9). In the exposed coalfield the Coal Measures are about 132 m thick, proved in the Dixon and Burne Colliery borehole. The Coal Measures in the South Staffordshire Coalfield thicken towards the north, but no complete section has been proved in the exposed coalfield of the district (Waters, 1991a, p.8). By comparison with sections to the west and east, it is probable that the now partly denuded Coal Measures (Figure 8) attained a thickness of about 200 m in the north of the district (compare with Whitehead and Eastwood, 1927, p.29; Hamblin et al., 1991). The most easterly proving in the concealed part of the coalfield is at Hamstead (Hamstead boreholes; Poole, 1970; Waters, 1993), where 199 m of Coal Measures are present in Hamstead No. 1 [0760 9625]. In the southern part of the district, the Coal Measures pass laterally into reddened and condensed successions; for instance, immediately east of the Russell's Hall Fault, a condensed Coal Measures succession of about 31 m is proved in Mucklow Hill Borehole.

Two regionally extensive marine bands the Vanderbeckei (Langsettian–Duckmantian boundary) and the Aegiranum (Duckmantian–Bolsovian boundary) are recognisable in places, in the north-west of the district (Figures 7 and 8). The former occurs locally above the Stinking Coal and is known as the Stinking

Table 13 Lithostratigraphical nomenclature of the Upper Carboniferous strata in the South Staffordshire and Warwickshire coalfields, compared to previous schemes (Besly, 1988; Old, 1987; 1990).

CHRONOSTRATIGRAPHY		LITHOSTRATIGRAPHY			
		South Staffordshire		Warwickshire	
		Besly, (1988)	This memoir (after Besly & Cleal, 1997)	This memoir (after Bridge et al. (1998))	Old et al. (1987; 1990)
Early Permian (?)		Clent Formation	Clent Formation	Kenilworth Sandstone Formation	Kenilworth Mudstone Formation*
? — ?		*(unconformity)*		Tile Hill Mudstone Formation*	Tile Hill Mudstone Formation
Late Carboniferous (Silesian)	Stephanian (?)	Enville Formation	Salop Formation — Enville Member	Meriden Formation — Allesley Member / Keresley Member — Enville Group	Coventry Sandstone Formation
		Keele Formation	Alveley Formation	Whiteacre Member	Keele Formation
	Westphalian D	Halesowen Formation			Halesowen Formation
	Bolsovian (Westphalian C) *Aegiranum MB*	Etruria Formation			Etruria Marl Formation
	Duckmantian (Westphalian B) *Vanderbeckei MB*	Middle Coal Measures			
	Langsettian (Westphalian A)	Lower Coal Measures			
	Namurian	*(unconformity)*		Millstone Grit (?)	

* does not crop out in district
MB—marine band
wavy lines indicate principal unconformities
vertical lines indicate significant erosion or depositional hiatus.

Marine Band. The latter occurs locally above the Two Foot Coal; to the west of the Great Barr Fault it appears to be restricted to the area north of the Gospel Oak and Ball's Hill faults (Figure 1).

LOWER COAL MEASURES

In this coalfield, the Vanderbeckei Marine Band is restricted in occurrence. It rests on the Stinking Coal, and in the absence of the marine band the coal is taken as the top of the Lower Coal Measures (Figure 7). In the south of the coalfield, where the Coal Measures are thinner and pass into locally reddened non-coal bearing strata only the Thick Coal can be recognised with any certainty south of Halesowen (Figure 8). About 70 to 84 m of Lower Coal Measures are present in the north (e.g. Hamstead No. 1 Borehole); this contrasts with over 100 m of Lower Coal

Figure 7 Schematic stratigraphical relationships of the Carboniferous rocks in the west of the district (South Staffordshire Coalfield). Not to scale. Vertical lines indicate main unconformities.

D dolerite intrusion; LCM Lower Coal Measures; MCM Middle Coal Measures; 2FT Two Foot Coal; S Stinking Coal; VMB Vanderbeckei Marine Band; AMB Aegiranum Marine Band; WBM Wales–Brabant Massif. (After Glover and Powell, 1996.)

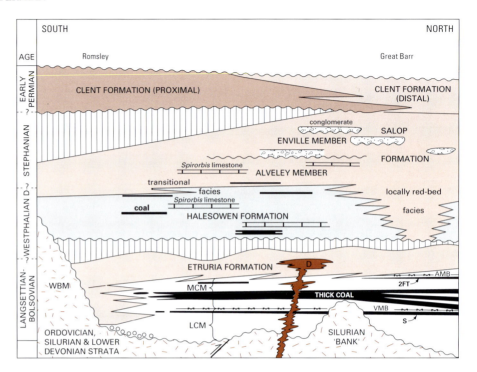

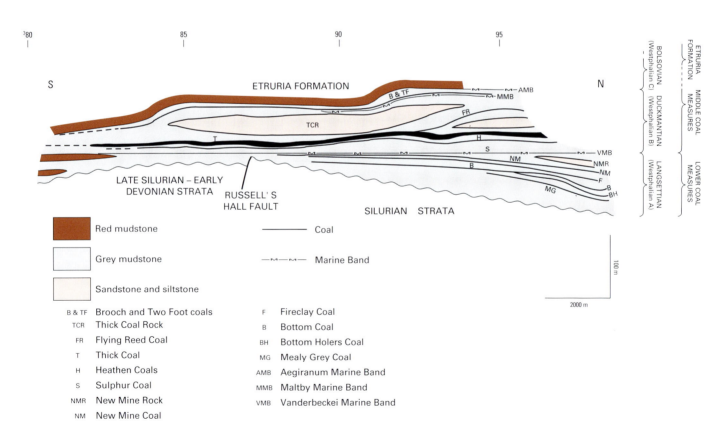

B & TF	Brooch and Two Foot coals	F	Fireclay Coal
TCR	Thick Coal Rock	B	Bottom Coal
FR	Flying Reed Coal	BH	Bottom Holers Coal
T	Thick Coal	MG	Mealy Grey Coal
H	Heathen Coals	AMB	Aegiranum Marine Band
S	Sulphur Coal	MMB	Maltby Marine Band
NMR	New Mine Rock	VMB	Vanderbeckei Marine Band
NM	New Mine Coal		

Figure 8 Generalised cross-section (from south to north) of the Coal Measures in the west of the district (South Staffordshire Coalfield).

Figure 9 Comparative vertical sections through the Coal Measures in the north-central and south of the South Staffordshire Coalfield.
Datum line is the Vanderbeckei (Stinking) Marine Band, above the Stinking Coal.

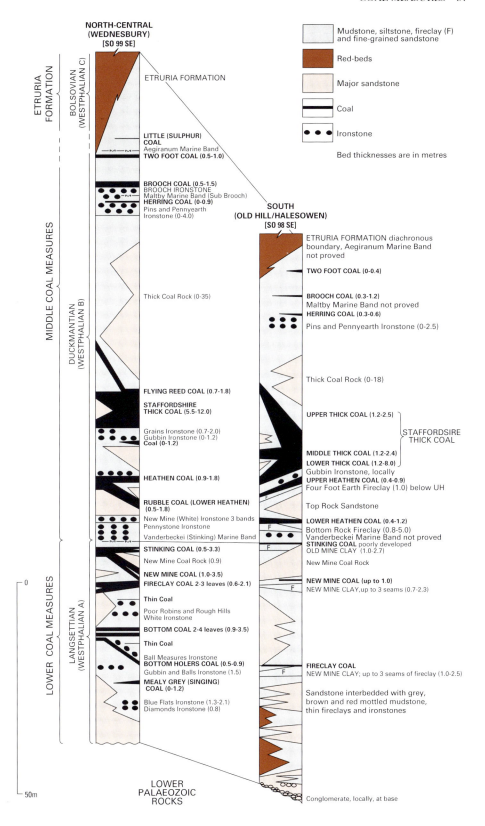

Measures strata which are present to the north of the district, north of the Bentley Trough Faults (Figure 8). Farther south, at Kate's Hill [9508 8994] and at the former site of Blakely Hall Colliery [999 895], the Lower Coal Measures are about 39 m thick; continued southward thinning is demonstrated on the east side of the Russell's Hall Fault by the presence of a condensed and apparently unfaulted succession, 11 m thick, proved at Old Blackheath Colliery [9760 8952]. The most southerly proving of Coal Measures in the district was at St Mary's Abbey, south of Halesowen, on the west side of the Russell's Hall Fault [976 831]. Here, about 60 m of mainly red mudstone and sandstone may be equivalent to the Lower Coal Measures (Figure 7), but the absence of well-developed coals beneath the Thick Coal make this figure tentative. This thickness is similar to that estimated for the Lower Coal Measures of the Wassell Grove Colliery [SO 934 820], to the west of the district (Whitehead and Eastwood, 1927, fig. 3; Poole, 1970).

The Lower Coal Measures consist of grey mudstone with subordinate sandstone. The mudstone is either laminated or exhibits pedogenic features such as root traces (rhizoliths), sphaerosiderite and siderite nodules and destratification. At the southern margin of the coalfield red and brown mudstones are also present (Figure 9). Sandstone beds occur most commonly towards the base of the succession (Whitehead and Eastwood, 1927; p.35, 44, 47), but nowhere are they as thick or as coarse and pebbly as those seen at the base of the Lower Coal Measures immediately to the west of the district (e.g. flanks of the Netherton Anticline or Dudley town). The only other notable development of sandstone in the Lower Coal Measures is a lens of sandstone termed the New Mine Coal Rock (Figure 9) which is thickest to the north of the Ball's Hill Fault. Locally, this sandstone rests directly on the New Mine Coal (Hamblin et al., 1992; Whitehead and Eastwood, 1927, p.42). Thin beds of coal up to about 2 m thick, which have previously been worked, are present throughout the succession across most of the area, but to the south many of the coal seams fail and these may pass laterally into rooted grey or pale grey mudstone (fireclay). In general, stratigraphically higher coals persist farther south (Figures 8, 9): the lowest seam, the **Mealy Grey (Singing Coal)** is present only in the far north of the district; the **Bottom Coal** and the **Fireclay Coal** can be traced south to about Northing 889; the **New Mine Coal** fails at about Northing 880, and the **Stinking Coal** fails at about Northing 845. Workable thicknesses of ironstones, normally in the form of sideritic nodules or thin beds, also occur sporadically. The best developed are the **Blue Flats Ironstone** and **Diamonds Ironstone** which occur beneath the Mealy Grey Coal in the north of the district (Figure 9). Several other named ironstone seams occur between the level of the Mealy Grey and the Fireclay Coal (Figure 9), but few well-developed ironstones occur between the latter and the Stinking Coal.

MIDDLE COAL MEASURES

The Middle Coal Measures of the South Staffordshire Coalfield varies in thickness from about 118 m in the north of the district (e.g. Tipton Green No. 6 Borehole) to about 50 m in the south. The top of the Middle Coal Measures is taken where red mudstone dominates the sequence. This boundary is clearly diachronous and gradational (Figures 7, 8) so that the thickness of the Middle Coal Measures can vary locally.

Like the Lower Coal Measures, the succession is dominated by grey mudstone with subordinate sandstone (Figure 9). Coal seams are present throughout the succession, and many of these have been worked extensively; the most persistent of these is the **Staffordshire Thick Coal**, but many of the other seams can also be traced over much of the coalfield. The Staffordshire Thick Coal, including thin interbeds of mudstone ('dirt' measures), is up to 12 m thick in this district. Up to 14 individual coal partings have been recognised within the Thick Coal 'group' (Whitehead and Eastwood, 1927; p.27). Each named coal parting has specific characteristics which enabled their identification over most of the coalfield. The **Flying Reed Coal,** which comprises the upper two of these 14 coals, splits northwards from the Thick Coal (Figures 8, 9) along a line approximating to the surface trace of the Tipton–Hill Top Fault [960 920 to 008 933] (Waters et al., 1994). The intervening measures are mainly sandstone (Whitehead and Eastwood, 1927, p.73); locally, coal has been removed by erosion and replaced by sandstone channel fills known as washouts. A further split in the Thick Coal is the result of a sandstone washout at Black Lake [990 942] (Figure 8); the coal at the base of the washout has been termed the **Chance Coal**. Other minor partings are present in the district, especially in the south where sandstone partings become more common. The remaining named seams in the coalfield are, in upwards sequence, the **Rubble (Lower Heathen), Heathen, Herring, Brooch** and the **Two Foot** coals (Figure 9). The Rubble Coal is a split from the main Heathen Coal, though the latter has, in places, been referred to as the Upper Heathen Coal. The 'Heathen coals' are present over much of this western area, but in places the coal has been replaced by sandstone washouts (e.g. Rowley Hall No. 9 Pit). However, south of Oldbury, the 'Heathen coals' lie close together or form a single seam. The Herring Coal, so-called because of the presence of dorsal spines of fish in the top of the coal (Jukes, 1859), occurs locally 1 to 4 m beneath the Brooch Coal, and attains a maximum thickness of about 0.6 m. The Brooch Coal is present across the entire South Staffordshire Coalfield; its average thickness is about 1.2 m, but it may be as much as 2 m thick. However, in Hamstead No. 1 Borehole, east of the exposed coalfield, the Brooch Coal is only 0.3 m thick and in the southern part of the district it is generally less than 0.5 m thick. The strata between the Brooch and Two Foot coals (Plate 4) is usually between 1 and 5 m thick. Exceptionally, beneath the former site of Round's Green Colliery, west of Oldbury, the parting is about 15 m and comprises mainly sandstone. The Two Foot Coal is present over much of the coalfield, but like so many of the underlying coals it becomes difficult to identify in the south and in the east (e.g. Sandwell Park No. 1 Borehole) where a 'black ring' (carbonaceous

Plate 4 View along the exhumed Coseley–Wednesbury Fault (looking east in 1990). Note the drag of the downthrown Coal Measures strata (left), including two thin coal seams, against the fault. The surface of the lower seam (Two Foot) in the foreground has been cleaned prior to excavation. Patent Shaft Opencast site [976 950] (GS 708).

mudstone) may be equivalent to the Two Foot Coal. A thin coal, termed the **Little Coal** or the **Sulphur Coal**, also occurs above the Two Foot Coal in the north of the district.

Several beds of ironstone (Figure 9) are thick enough to have been worked in the past. Beds of nodular sideritic ironstone known as the **Pennystone** occur above the Stinking Coal over much of the district. The most extensively worked ironstone, the **New Mine (White)**, occurs beneath the Rubble (Lower Heathen) Coal, and is up to 3 m thick in the north of the district. Laterally persistent beds of nodular or thinly bedded sideritic ironstone also occur between the Heathen and Thick coals (Lamb/ Grains and Gubbin ironstones) and beneath the Herring Coal (Pins and Pennyearth Ironstone; Figure 9). A laterally impersistent ironstone bed, the Brooch Ironstone, occurs directly beneath the Brooch Coal.

The main development of sandstone and siltstone in the Middle Coal Measures in this western area of the district is the Thick Coal Rock (Figures 8, 9). This term was used specifically for a coarse-grained, locally pebbly sandstone, about 30 m thick, above the Thick Coal (including seams equivalent to the Flying Reed Coal) in the vicinity of the Dudley Port Trough faults (Whitehead and Eastwood, 1927, p.73). Towards the north and south this sandstone unit becomes interspersed with mudstone partings and it is generally finer grained (Hamblin and Glover, 1991b, fig. 4).

The **Vanderbeckei (Stinking) Marine Band** has been proved at Kate's Hill and Oldbury (Whitehead and Eastwood, 1927; p.45; Jukes, 1859; p.58). The marine band consists of dark grey mudstone with pyritised patches; in Kate's Hill No. 2 Borehole it is about 1.3 m thick and contains fossiliferous ironstone nodules. Jukes (1859) lists a number of marine and brackish-water fossils recovered from the Pennystone Measures (directly overlying the Stinking Coal) of the Oldbury–Portway area [975 885]. These include *Productus scabricula*, *Anthracosia*, *Avicula quadrata*, *Pecten* (?), *Lingula mytiloides* (?), *Orbicula nitida*, *Conularia quadrisulcata* and fish teeth and bones.

The **Maltby Marine Band** is equated with a fossiliferous, black mudstone succession which lies directly beneath the Brooch Coal or Ironstone (Figures 8, 9); this is locally termed the Sub-Brooch Marine Band and is well developed in the north of this district, but it has not been identified south of Northing 890 (Oldbury). This marine band, or more correctly *Lingula* band, was exposed at Patent Shafts opencast site [977 949] during the time of resurvey. At this locality it consists of about 1 m of black and dark grey, slightly fissile mudstone, rich in compressed nonmarine bivalves (*Anthraconauta?* and *Anthracosia?*) and *Lingula* sp.

The **Aegiranum Marine Band** (Figures 8, 9), locally known as the Charles or Hamstead Marine Band is poorly developed about 2 m above the Two Foot Coal in the Kings Hill area [98 95] but is largely absent to the south of the Gospel Oak–Ball's Hill faults. Provings at Sandwell Park and Hamstead collieries, as well as in the Hamstead boreholes (e.g. Hamstead No. 1), record a *Lingula* band above the Brooch Coal which is believed to be equivalent to the Aegiranum Marine Band (Poole, 1970; Whitehead and Eastwood, 1927; p.32).

Warwickshire Coalfield

The Warwickshire Coalfield forms the eastern margin of the sheet, and is bounded to the west by the north-trending Warwickshire Western Boundary Fault. Coal Measures do not crop out, but the sequence (Figures 10, 11) is well known from mine workings at Dexter Colliery [2436 9502] (Figure 12) in the north-east of the district, at Daw Mill (Coventry district) and Kingsbury (Lichfield district) collieries, and from shaft and borehole records (Chapter Ten).

The stratigraphy and depositional environments of the Coal Measures in the Warwickshire Coalfield are fully described elsewhere (Eastwood et al., 1923, 1925; Mitchell, 1942; Fulton and Williams, 1988; Fulton and Guion, 1990; Jones, 1992; Bridge et al., 1998). Consequently, only a brief review is given here, together with details of the sequence proved in this district (Figure 11; Table 13). The maximum thickness of the Coal Measures in this district is about 158 m, proved in the Whitehouse Farm Borehole; this compares with an average regional thickness of about 111 m in the south-central part of the coalfield (Old, 1989) and about 160 m to the east of the district (Bridge et al., 1998).

Boreholes in the district and the adjacent coalfield area to the east proved Westphalian Coal Measures resting unconformably on late Cambrian or early Ordovician strata (Figure 11), comprising the Monks Park Shale and Merevale Shale formations (Chapter Three). However, a lensoid pebbly sandstone (about 25 m thick) with subordinate *Lingula*-bearing mudstone which crops out locally near Dost Hill in the Lichfield district, has been tentatively assigned to the Millstone Grit of Namurian age (Taylor and Rushton, 1971). The Vanderbeckei Marine Band, where it has been recognised, lies about 5 m below the base of the Thick Coal (Figures 10, 11), and marks the boundary between Langsettian and Duckmantian strata. The Maltby (Four Feet) Marine Band, which lies above the Four Feet Coal, is present locally (Figure 11). The Aegiranum Marine Band, which marks the base of the Bolsovian, has not been recognised in this district (Figure 11), probably due to the lateral passage into continental red-bed lithofacies (Etruria Formation) (Fulton and Williams, 1988).

LOWER COAL MEASURES

Lower Coal Measures strata below the Vanderbeckei Marine Band typically range in thickness from 30 to 50 m (Figure 11); the thin succession proved in the Priory Wood Borehole is anomalous and probably due to local faulting. The Whitehouse Farm Borehole, located close to the Western Boundary Fault, however, proved 86 m of these beds (allowing for a reported dip of about 20°).

The strata consist mainly of grey mudstone, seatearth mudstone (fireclay), and thin sandstone beds, although in the Outwoods Borehole [2462 8528] the proportion of sandstone is greater (Figure 11). Palaeosols (seatearths) include partially drained, sphaerosiderite-rich, brown, red, purple and yellow mudstone with sparse carbonaceous rootlet traces; poorly drained palaeosols, consisting of fine- or coarse-grained sediment, are characterised by grey or dark grey colours, and are rich in carbonaceous rootlet traces and organic matter; siderite nodules are common. Thin coals occur in most of the boreholes, but correlation of unnamed seams has not been possible, except for underground boreholes at Kingsbury [2402 9354; 2470 9373] in the north-east of the district (Figure 10), and in the Whitehouse Farm Borehole and Dove House Farm Borehole [2472 8912] located farther the south (Figure 11). The following named seams, in upward sequence, can be correlated over the district (Figures 10, 11): **Stanhope, Stumpy, Bench, Top Bench, Doubles, Deep Rider, Yard and Seven Feet**.

Sparse data in the Kingsbury area indicate a local trend of westward thickening of the Lower Coal Measures towards the Western Boundary Fault, accompanied by an increase in both the number and thickness of the seams (Figures 10, 11). The seams generally are less than a metre thick, except for the **Top Bench** and **Bench** coals which are 1.52 m and 1.57 m thick, respectively, in the Kingsbury No. 3 Underground Borehole (Figure 12). In contrast to this local trend of westward thickening, regional interseam isopachyte maps for the Lower Coal Measures (Mitchell, 1942; Bridge et al., 1998) show a trend of increasing thickness (reflecting increased subsidence) towards the north-east, from about 30 m in the Outwoods Borehole, to about 160 m in the north-east of the coalfield.

MIDDLE COAL MEASURES

The Middle Coal Measures succession above the Vanderbeckei Marine Band typically ranges in thickness from 55 to 60 m (Figure 11), but the Whitehouse Farm Borehole, located close to the Western Boundary Fault, proved about 72 m (allowing for a reported dip of 20°). The Middle Coal Measures are lithologically similar to the Lower Coal Measures, but there is a higher proportion of sandstone beds, commonly about 2 m thick; channel-fill sandstones (washouts) are locally present.

A thick, geographically extensive washout, about 1 km wide, and trending north-north-west, cuts out the Two Yard Coal and underlying strata down to the level of the Ryder Coal over much of the area of the Kingsbury and Dexter collieries (Bridge et al., 1998); up to 20 m of channel-fill sedimentary rocks were recorded in the Kingsbury (Dexter) No. 3 Shaft (Figure 12), but the main sandstone is about 6 m thick (Guion and Fulton, 1986). The washout forms part of a larger channel belt, up to 1.2 km wide; palaeocurrent measurements along the Dexter Manrider Roadway (Guion and Fulton, 1986; Jones, 1992) indicate palaeoflow to the south-east. The erosive base of the washout is locally overlain by an intraformational mudstone/coal clast conglomerate, which passes up through an upward-fining sequence, from sandstone to interbedded sandstone and siltstone. The lower part of the channel sandstone is characterised by epsilon cross-bedding, typical of point-bar deposits of meandering stream channels (Allen, 1965), passing up to trough cross-laminated sandstone-siltstone beds with climbing ripples and *Pelecypodichnus* sp. bivalve escape structures (Guion and Fulton, 1986). The channel-fill

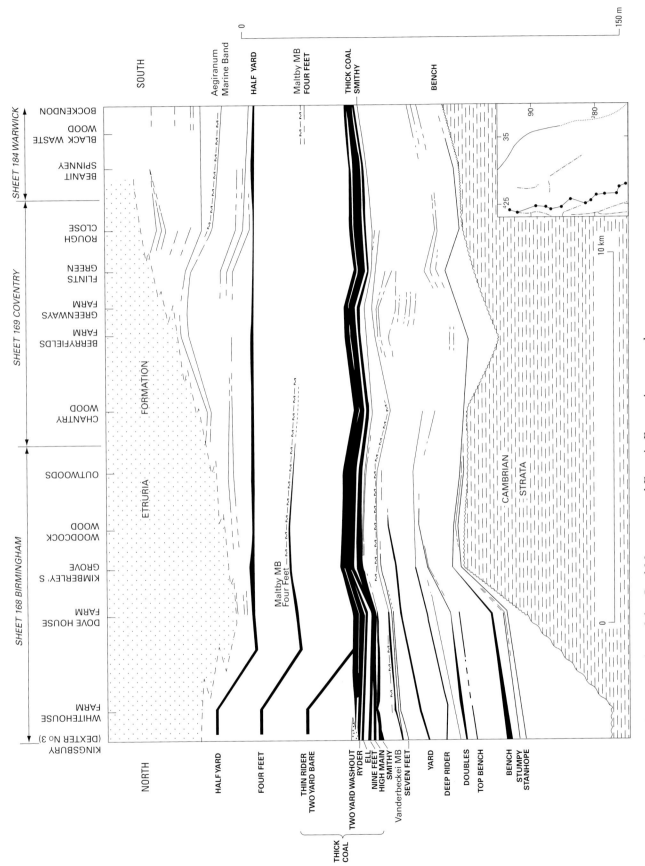

Figure 10 North–south cross-section of the Coal Measures and Etruria Formation near the western margin of the Warwickshire Coalfield, showing correlation of the coal seams and marine bands. Inset map shows location of boreholes. Adapted from Jones (1992).

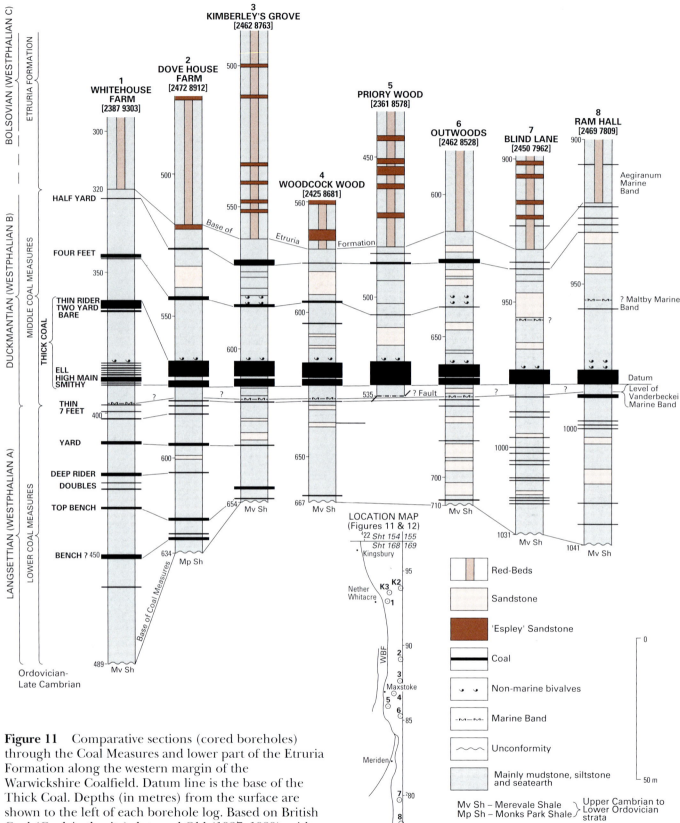

Figure 11 Comparative sections (cored boreholes) through the Coal Measures and lower part of the Etruria Formation along the western margin of the Warwickshire Coalfield. Datum line is the base of the Thick Coal. Depths (in metres) from the surface are shown to the left of each borehole log. Based on British Coal (Coal Authority) data and Old (1987, 1989), with additions. Location map shows the position of boreholes 1 to 8 (this figure) and underground boreholes K2 and K3 (Figure 12) in relation to the eastern boundary of the district. WBF Western Boundary Fault.

Figure 12 Graphic logs of the Coal Measures in the Kingsbury (Dexter) underground boreholes No. 2 and No. 3, showing the principal coal seams. Depths are in metres below OD. Datum line is the Vanderbeckei Marine Band. See Figure 11 for location map (K2 and K3).

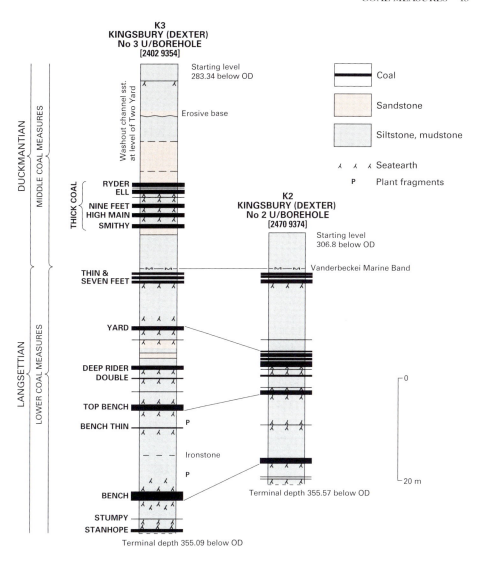

sandstone was formerly exposed about 3 km north of the district near Dost Hill.

The **Vanderbeckei Marine Band**, locally termed the Seven Feet Marine Band, consists of dark grey, locally pyritous mudstone, commonly with ironstone nodules; fauna include the brachiopods *Dunbarella* sp., abundant *Lingula mytiloides* (juv.) and *Lingula* sp., 'mussels', fish remains and ostracods. In the Woodcock Wood Borehole, the shales with *Lingula* sp. are overlain by a thin ironstone which yielded *Spirorbis* sp. and the bivalve *Anthracosia* sp.

The **Thick Coal** lies 3 to 7 m above the Vanderbeckei Marine Band (Figures 11, 12) and represents the only seam of economic importance in the district. The Thick Coal 'group', including 'dirt bands' (siliciclastic laminae and thin beds), ranges in thickness from 6 to 8.5 m. In the north-east of the district, it is represented by up to eight individual named seams (Figure 11), separated by siliciclastic rocks; the seams, in ascending order, are the **Smithy, High Main, Nine Feet, Ell, Ryder, Bare, Two Yard and Thin Rider** (Cope and Jones, 1970; Old et al., 1990; Bridge et al., 1998). Fulton (1987) excludes the

Smithy Coal from the Thick Coal 'group'. Over much of the coalfield the constituent seams are amalgamated, since the 'dirt bands' are extremely thin (generally less than 0.3 m thick), although all the seams were recognisable in the Dove House Farm Borehole, with the only significant 'dirt band' occurring between the Nine Feet and Ell coals (Cope and Jones, 1970). However, thicker interseam beds were encountered in the north-east of the area in the Kingsbury No. 3 Underground Borehole (Figure 12). Regional isopachyte trends for the Thick Coal 'group' (Mitchell, 1942; Bridge et al., 1998) indicate an area of amalgamated seams in the south-west of the district, passing north-eastward with increasing interseam thickness, from about 8 m at Dexter Pit [SP 2436 9502] to 15 m at Kingsbury Pit [SP 2296 9821] on the adjacent sheet to the north. Another zone of split Thick Coal occurs adjacent to the Western Boundary Fault in the north-east of the district (Bridge et al., 1998, fig. 22). Flora identified from this internval includes *Neuropteris* sp., scattered *Stigmaria* roots, and sparse in-situ tree trunks (Guion and Fielding, 1988); mudstone immediately above the Thick Coal commonly contains non-marine bivalves including *Anthracosia* sp., and *Naidites* sp.

Other named coals, correlatable throughout most of the coalfield are, in upward sequence, the **Four Feet** and the **Half Yard** (Figure 11). The former reaches a maximum thickness of 1.1 m in the Dove House Farm Borehole. Overlying mudstones contain nonmarine bivalves and fish scales, and in the Woodcock Farm Borehole they also yielded *Lingula* sp., probably representing the Maltby Marine Band. This marine band may also have been proved farther south in the Ram Hall and Blind Lane boreholes (Figure 11), where bivalves, including *Lingula* sp. and a productid, and fish fragments were reported by British Coal. Three seams (0.11, 0.64, and 2.58 m thick, in upward sequence) at about 363 m depth in the Whitehouse Farm Borehole (Figure 11) may represent the Four Feet Coal, although British Coal correlated these seams as splits of the Thick Coal (Ryder, Bare and Thin Ryder/Two Yard), in upward sequence. The Half Yard Coal reaches a maximum thickness at Kimberley's Grove, where it comprises upper and lower leaves, 0.67 and 0.92 m thick, respectively, separated by 0.3 m of mudstone. Interseam isopachyte maps for the strata between the top of the Thick Coal 'group' and the base of the Etruria Formation show a regional trend of increasing thickness towards the north-east (Mitchell, 1942; Bridge et al., 1998).

The strata between the Two Yard Coal and the base of the red-bed Etruria Formation and the (Figure 11) are not well known in the northern part of the district since underground colliery boreholes penetrating the upper part of the workable Coal Measures generally commenced in workings of the Two Yard Seam (Figure 12).

The upper boundary with the red-bed Etruria Formation is both diachronous and gradational (Mitchell, 1942; Jones,1992). In this district, it lies about 5 to 10 m above the Half Yard Coal (Figure 11), and is taken at the incoming of persistent reddened mudstones (oxidised palaeosols) together with coarse-grained sandstones and conglomerates ('espleys').

Depositional environments and sedimentology

The Coal Measures within the Warwickshire Coalfield and the South Staffordshire Coalfield were deposited close to the southern margin of the Pennine Basin, adjacent to the Wales Brabant Massif. This massif comprised a slowly subsiding hinterland block which had a significant effect on Westphalian sedimentation, as Coal Measures sediments onlapped southwards.

Eight sedimentary facies can be recognised in the Coal Measures (Guion et al., 1995; Jones, 1992): lacustrine, lacustrine delta, distributary channel, overbank and crevasse splay, palaeosol, mire (coal) and marine. These facies, except for the last, formed in a low-lying alluvial plain setting characterised by extensive, shallow, freshwater lakes. Lacustrine and lacustrine delta facies consisting predominantly of mudstones and siltstones — many with rootlet traces — form the largest proportion of the Coal Measures succession. Lakes were infilled by sediment from suspension and by tractionally deposited bedload fed by lacustrine delta, crevasse splay and overbank deposits which were in places associated with highly sinuous, laterally switching, distributary channels (e.g. Thick Coal Rock) including washouts. Regional-scale abandonment and infilling of the shallow lakes resulted in extensive colonisation by plants and the development of gley (waterlogged or hydromorphic) palaeosols; continued low rates of subsidence and siliciclastic input, high rates of precipitation and repeated flooding resulted in the development of low lying-mires (rheotrophic swamps), and, in the case of thicker accumulations such as the Warwickshire and Staffordshire Thick coals, raised mires (ombrotrophic bogs). Subsequent renewed subsidence resulted in drowning of the mire, the development of interdistributary lakes, and the initiation of another lake-filling cycle.

In the southerly areas of the South Staffordshire Coalfield, the former margins of the Pennine Basin, coal-bearing strata pass laterally into stacked palaeosol successions. This area was fed sporadically with siliciclastic material which inhibited peat formation. Locally, the water table dropped sufficiently for oxidising processes and siderite formation to take place; the former led to red and brown mottling of the mudstones (semi-gley and brunified alluvial palaeosols), and the latter process led to growth of sphaerosiderite and larger siderite nodules. These seatearth mudstones or fireclays were the basis of a major extraction industry in the early part of the 20th century; similarly the widespread development of sideritic mudstones through both these pedogenic processes and eogenetic processes in the lacustrine mudstones formed the basis of widespread ironstone extraction in the South Staffordshire Coalfield.

The Vanderbeckei Marine Band marks a flooding of the entire basin. It is characterised by thinly laminated dark grey to black mudstone, with a fully marine fauna over much of the basin. The absence of the Aegiranum Marine Band over most of the district reflects the progressive northwards progradation of better drained alluvial conditions.

In the South Staffordshire Coalfield, a series of boreholes in the Great Bridge area document the lateral and vertical transition between the Coal Measures and the Etruria Formation (Hamblin and Glover, 1991; Figure 13, 14). The Little Coal, which caps the Coal Measures in this area is laterally impersistent, and there is a suggestion of a catenary relationship between the Little Coal, brunified alluvial palaeosols and lacustrine deposits across the Gospel Oak Fault (Figure 13).

A number of factors influenced sedimentation during deposition of the Coal Measures. The principal allocyclic controls were climate, subsidence rates, tectonics, and glacio-eustasy; autocyclic controls include sediment compaction and sedimentary processes such as channel switching.

Central England occupied an equatorial position during deposition of the Coal Measures (Scotese and McKerrow, 1990), and factors such as prolific vegetation indicate a humid tropical climate, probably with little seasonal variation (Scott, 1979).

Differential subsidence is reflected in a thickening of the succession to the north of the district, towards the centre of the Pennine Basin (Figure 8). Onlap of Langsettian and Duckmantian strata on to the Wales–Brabant Massif points to progressive enlargement of the Pennine Basin as sedimentation kept pace with subsidence. Differential rates of subsidence, from north to south, controlled the amalgamation (low rate of subsidence) and splitting (high rate of subsidence) of the Thick Coal. The trend of reduced subsidence in the south of the district is also reflected in the southward lateral transition from marine to brackish faunas in the marine bands (Fulton, 1987b), and also in the southward lateral passage from coal-rich strata to stacked, gley, semigley and brunified alluvial palaeosols in the South Staffordshire Coalfield.

Local tectonic controls may have been important. In the Warwickshire Coalfield, the north–south zone of splitting within the Thick Coal runs subparallel to the Western Boundary Fault. The abrupt northward thickening of Langsettian strata between Kimberley's Grove and Whitehouse Farm Boreholes (Figure 11) indicates differential subsidence in this area, possibly controlled by contemporaneous movement along faults in the basement rocks. In the South Staffordshire Coalfield, splits in the Thick Coal, such as the Flying Reed, appear to coincide with fault trends (Waters et al., 1994). Lateral facies relationships at the Coal Measures–Etruria Formation boundary also suggest possible fault influences (Figure 13, Little Coal across the Gospel Oak Fault; Waters et al., 1994, fig. 11).

Glacioeustatic controls are manifested in the basinwide, marine flooding sequences represented by the marine bands and their equivalent *Lingula* bands. These represent a rapid and geographically widespread rise in relative sea level and have been linked to variations in global climate which resulted in fluctuations in the polar ice-caps in Westphalian times (Leeder, 1988; Collier et al., 1990).

It is difficult to determine the effects of differential sediment compaction from other factors. In the Warwickshire Coalfield, Fulton and Williams (1988) could not establish a relationship between the area of amalgamated Thick Coal (likely to be subject to increased compaction) and thicker overlying interseam sediments. In the South Staffordshire Coalfield, however, the Thick Coal Rock shows a clear offset stacking pattern with the split between the Flying Reed and the remainder of the Thick Coal (Figure 8). Sedimentary controls relate to variations in the supply and dispersal of sediment, and include channel switching and avulsion as a result of the enlargement and variations in river discharge, and variations in geomorphological gradient. These autocyclic processes were probably the cause of much lacustrine and channel infilling.

ETRURIA FORMATION

The Etruria Formation transitionally overlies the Coal Measures in both the Warwickshire and Staffordshire coalfields (Figures 8, 11). The base is taken where the grey beds pass up into predominantly red mudstones (Whitehead and Pocock, 1947, p.59; Hoare, 1959). Thickness variations are large, due to the diachronous boundary with the Coal Measures, and also due to the influence of fault movements and erosion prior to deposition of the overlying Halesowen Formation. The Etruria Formation consists predominantly of mudstone which is mainly red, but which may also exhibit a wide variety of colours and mottles such as brown, green and yellow; this distinctive coloration is due mainly to subaerial, oxidising pedogenesis prior to sediment burial (see below). Thin coals or dark grey carbonaceous beds are developed, locally, at the transitional lower boundary with the Coal Measures (Figures 13, 14). Subordinate amounts of sandstone and conglomerate, locally known as espleys, are also present throughout the formation. Extraformational pebbles indicate a wide variety of provenance, but most are believed to be derived from the Wales–Brabant Massif or local highs adjacent to the southern margin of the Pennine Basin (see below).

South Staffordshire Coalfield

The Etruria Formation, formerly termed the Old Hill or Oldbury Marls (Jukes, 1859; Kay, 1913), crops out over much of the southern part of the South Staffordshire Coalfield in this district. In the north, its outcrop is bounded by two east-trending faults, the Coseley–Wednesbury Fault and the Ball's Hill Fault, so as to form a stepped outcrop pattern juxtaposed against the Coal Measures to the north. To the east it is faulted against the Salop Formation, and in the south it is overlain unconformably by the Halesowen Formation.

The maximum proved thickness of the formation is about 250 m (e.g. Yew Tree Colliery [9667 8678]), although the top of the formation is not present hereabouts, so that the original total thickness of the formation is likely to have been greater than this. There is great variation in the thickness of the formation due mainly to the unconformable nature of the contact with the overlying Halesowen Formation. Local structural movements and subsequent erosion prior to deposition of the Halesowen Formation (see Chapter Nine) have led to a thinner succession (about 30 m) being preserved to the east of the Russell's Halls Fault and in the concealed part of the South Staffordshire Coalfield, to the east of the Great Barr Fault.

The sandstones and conglomerates of the Etruria Formation, in this part of the district, are mostly poorly sorted and contain a high proportion of chloritic clays which gives them a characteristic green coloration where fresh, or an orange-brown coloration where weathered. Many of the clasts are also chloritic, reflecting weathering and derivation from basic igneous source rocks. Other clast types include white and pink quartzite, tuff, lapilli tuff, porphyritic acid igneous rock and sandstone, as well as intraformational mudstone clasts. Large pebble-size and coarser clasts are mainly subrounded to well rounded, but smaller clasts are mostly subangular to subrounded. Some of the coarsest conglomerates are

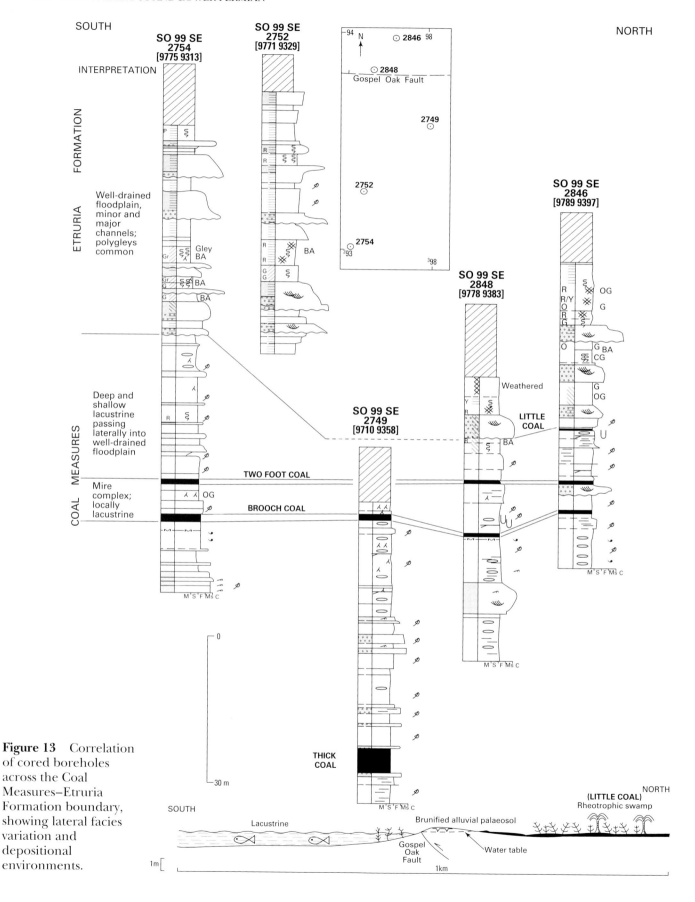

Figure 13 Correlation of cored boreholes across the Coal Measures–Etruria Formation boundary, showing lateral facies variation and depositional environments.

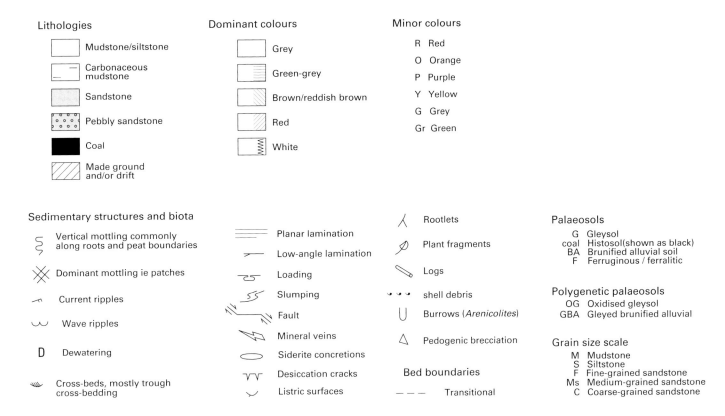

Lithologies

☐ Mudstone/siltstone

☐ Carbonaceous mudstone

▦ Sandstone

⊡ Pebbly sandstone

■ Coal

▨ Made ground and/or drift

Dominant colours

☐ Grey

☐ Green-grey

▨ Brown/reddish brown

▨ Red

≋ White

Minor colours

R Red
O Orange
P Purple
Y Yellow
G Grey
Gr Green

Sedimentary structures and biota

§ Vertical mottling commonly along roots and peat boundaries

✕ Dominant mottling ie patches

⌐ Current ripples

∪ Wave ripples

D Dewatering

⩘ Cross-beds, mostly trough cross-bedding

═ Planar lamination

⌐ Low-angle lamination

⊂⊃ Loading

∫ Slumping

⋀ Fault

⚡ Mineral veins

⬭ Siderite concretions

⊤⊤ Desiccation cracks

∪ Listric surfaces

⋀ Rootlets

⌀ Plant fragments

⟍ Logs

••• shell debris

∪ Burrows (*Arenicolites*)

△ Pedogenic brecciation

Bed boundaries

– – – Transitional

Palaeosols

G   Gleysol
coal Histosol(shown as black)
BA   Brunified alluvial soil
F   Ferruginous / ferralitic

Polygenetic palaeosols

OG   Oxidised gleysol
GBA Gleyed brunified alluvial

Grain size scale

M Mudstone
S Siltstone
F Fine-grained sandstone
Ms Medium-grained sandstone
C Coarse-grained sandstone

Figure 13 *continued.* Key for figures 13, 14 and 15.

exposed at Coombeswood [9724 8522] where cobbles and boulders of red and brown quartzite are present in a clast-supported conglomerate. Conglomerate and pebbly sandstone beds are present throughout the formation as is demonstrated by small exposures along the River Stour in the north-east of Halesowen and in the Haden Cross area where these lithologies form prominent ridges (Glover, 1990a; Waters, 1991b). These beds are commonly several metres thick and persist laterally for 500 to 900 m, but some can be traced over a few kilometres. Cored boreholes from the Great Bridge area demonstrate the complexity of colour mottling present in the mudstone of the Etruria Formation (Hamblin and Glover, 1991b; Figures 13, 14, 15); there is much evidence of plant colonisation, but also of desiccation and resultant brecciation of the sediment surface prior to burial (see below).

Warwickshire Coalfield

In the east of the district the Etruria Formation has been proved in a number of boreholes and shaft sections. The formation crops out about 2 km north of the district near Dost Hill where it is well exposed at Cliff Quarry [220 988]. The Etruria Formation consists predominantly of red mudstone with less common purple, green, grey, yellow and brown mudstone and subordinate coarse-grained, pebbly sandstone and conglomerates. The coarse-grained rocks are locally known as espleys and are typically laterally impersistent. In this area, the clasts

commonly comprise small, subangular, discoidal fragments of pale grey and green mudstone and siltstone, locally derived from the Cambrian and Ordovician basement rocks, together with quartzite pebbles and lithic fragments.

The Etruria Formation, proved in boreholes, ranges in thickness from 90 to 150 m in the Warwickshire Coalfield. Isopachyte maps indicate a trend of thickening south-westwards towards the Western Boundary Fault (Mitchell, 1942; Bridge et al., 1998), and this trend is, in part, due to lithofacies changes. Borehole data (Figure 10) indicate that the onset of reddening occurred at a lower stratigraphical level in the south-west of the coalfield, a few metres above the Half Yard Seam, and the reddening passes diachronously north-eastwards and south-eastwards to higher stratigraphical levels, about 25 to 30 m above the Aegiranum Marine Band. The presence of an unconformity at the base of the overlying Halesowen Formation may also account, in part, for lateral variations in thickness of the Etruria Formation.

Depositional environments and sedimentology

The Etruria Formation in this district comprises a well-drained, alluvial floodplain facies association (Besly and Fielding, 1989; Glover et al., 1993); in addition, a volcanic facies association, similar to that described from outcrops in the South Staffordshire Coalfield (Besley and Fielding, 1989; Glover et al., 1993), has been proved in boreholes in the Coventry district (Sheet 169) (Bridge et

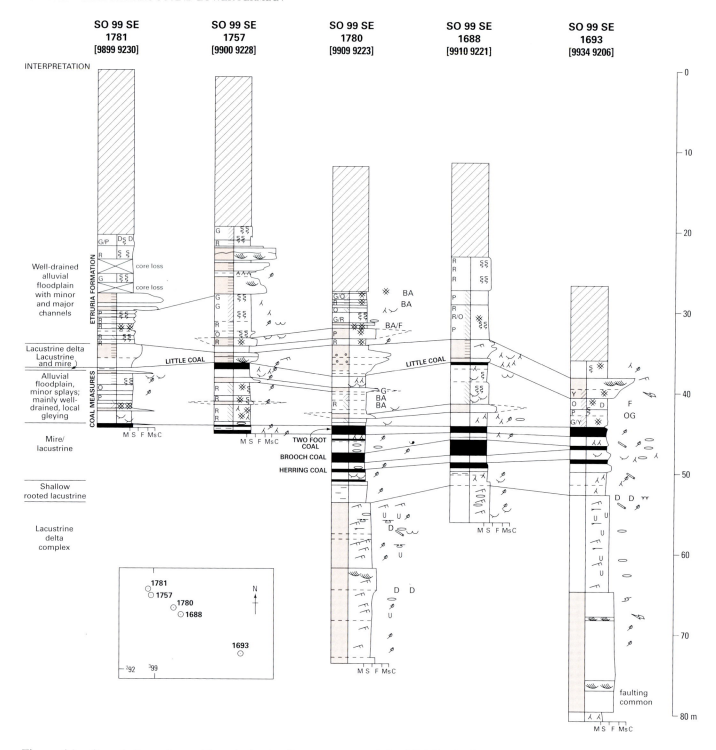

Figure 14 Correlation of cored boreholes in the upper part of the Middle Coal Measures and lower part of the Etruria Formation in the Great Bridge area, showing lateral facies variation and depositional environments. See p.47 for key.

al., 1998). The well-drained, alluvial floodplain association comprises red, purple, yellow and brown mudstones (brunified palaeosols) which, near the base of the formation, are interbedded with a few grey and carbonaceous mudstones and inferior coals reflecting the interdigitation of poorly drained and well-drained conditions near the boundary between the Coal Measures and the Etruria Formation. Well-developed palaeosol profiles, termed ferruginous and ferallitic palaeosols, which are indicative of lengthy periods of subaerial exposure and oxidation, are also present in places (Figure 14). Most sandstones represent either a single channel fill or multilateral, multistorey channel fills; proximal alluvial fan deposits are present in places.

The lower 40 m of the Etruria Formation in the Great Bridge area [98 92] of the South Staffordshire Coalfield consists of alternating brunified alluvial palaeosols and gley and polygenetic gley palaeosols (i.e. previously oxidised prior to gleying) (Figure 13, 14). Pebbly 'espley' sandstones and non-pebbly sandstones form single storey and multistorey channels. Finer-grained sandstones and siltstones occur mainly as lacustrine delta deposits, which can be traced locally over hundreds of metres (Figure 14). This periodically waterlogged lithofacies association has also been recognised in the lower part of the formation in the Tansey Green area [SP 91 89], a few kilometres west of the district (Glover et al., 1993). Above this transitional part of the Etruria Formation, red mudstones become more common and pedogenic features characteristic of well-drained conditions are better developed and more commonplace (e.g. desiccation cracks, ferruginous and ferralitic palaeosols). Minor waterlogged successions are also present but these tend to be in association with abandoned channels (e.g. Borehole 2115, Figure 15).

Exposures in Cliff Quarry [SP 220 988] located a few kilometres to the north in the Lichfield district (Sheet 154) show a predominantly mudstone sequence with sparse 'winged' channels. These are infilled with pebbly, coarse-grained sandstone and conglomerate, and lenticular pebbly mudstone and muddy pebble conglomerate. Clasts mostly comprise disc-shaped, subangular, pale grey-green, indurated mudstone and siltstone, probably locally derived from the Lower Palaeozoic Stockingford Shale Group (Barrow et al., 1919; Besly, 1988), which is interpreted to have formed a palaeohigh, located to the west of the Western Boundary Fault, during deposition of the Etruria Formation. Other clasts include small, well-rounded vein quartz pebbles and volcanic clasts. The majority of this association was deposited as overbank deposits on a well-drained alluvial plain, dominated by brunified palaeosols. The presence of sphaerosiderite, sparse plant fragments and grey-yellow mottled mudstones (interpreted as semi-gley palaeosols) indicate the intermittent, local development of poorly drained, waterlogged conditions on the floodplain. The isolated 'winged' channels are interpreted as the product of high sediment flux discharging within narrow, 'cut-and-fill', low-sinuosity channels, which were locally sourced from the Lower Palaeozoic hinterland. The pebbly mudstones and muddy pebble conglomerates are the products of locally derived, minor debris flows.

Proximal alluvial fan facies have been described from boreholes (e.g. Priory Wood and Outwoods) located adjacent to the western margin of the coalfield, in a zone within 3 km of the Western Boundary Fault (Besly, 1988; Bridge et al., 1998). However, the sparsity of mass flow deposits at Cliff Brickworks, described above, suggests that the coarse-grained and conglomeratic sandstones, whilst probably derived from an eroding hinterland source, were predominantly transported across the floodplain as channelised, traction-dominated, bedload deposits.

HALESOWEN FORMATION

The Halesowen Formation lies unconformably on the Etruria Formation and crops out in both the eastern and western areas of the district. The type area at Halesowen [967 866] within the district includes sections through the lower parts of the formation at The Rumbow [967 836] and along Illey Brook [975 816 to 977 813]. It consists of thick beds of pale greenish grey, fine- to coarse-grained, micaceous, locally pebbly sandstone, interbedded with grey mudstone, siltstone, and a few thin coals; thin beds of pale grey, micritic limestone (*Spirorbis* limestone) containing the annelid *Spirorbis* sp. are also present. Orange-brown nodules of calcite associated with caliche palaeosol formation are locally present in the mudstone beds, and also as reworked clasts in the coarse-grained sandstone beds. Thin red mudstone beds are especially common towards the top of the formation. Across most of the district the formation generally consists of three sandstone bodies separated by variable thicknesses of mainly grey mudstone with thin beds of coal and *Spirorbis* limestones. The uppermost sandstone is overlain by grey and red mudstones which pass upwards into the overlying Salop Formation (South Staffordshire Coalfield) and Meriden Formation (Warwickshire Coalfield). The average thickness of the formation is about 100 m.

Sandstones in the Halesowen Formation are petrologically distinctive; most are litharenites (McBride, 1963) with framework grains largely comprising monocrystalline quartz and phyllitic rock fragments rich in mica (chlorite, muscovite, and biotite) and some polycrystalline quartz. Leached sodic feldspars and, less commonly, potassic feldspars are also present (Lott, 1992a, b; Glover and Powell, 1996). Cements comprise ferroan spar and euhedral ferroan dolomite, together with fine kaolinite infilling pores. The heavy mineral suite is diverse. The presence of chloritoid, chrome spinel and abundant low pyrope garnets with variable grossular and spessartine contents indicates a source comprising low- to moderate-grade metasediments, probable granites and some ultramafics (Hallsworth, 1992a, b).

Palynological determinations from the Halesowen Formation in the district and surrounding areas (Owens in Hamblin, 1984; McNestry, 1994; Turner, 1994) indicate a Westphalian D age. Samples of the formation collected from Cliff Quarry, [219 984], near Dost Hill, in

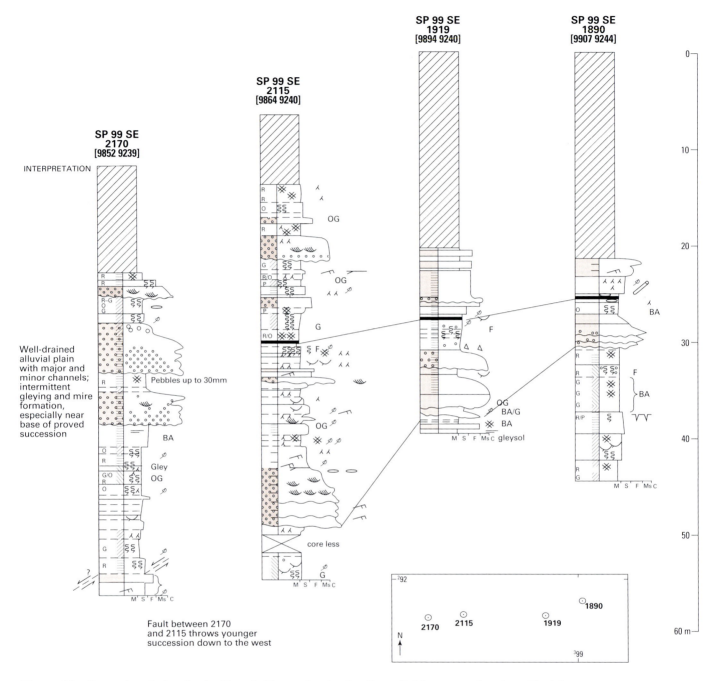

Figure 15   Lateral variation in the Etruria Formation in the Great Bridge area, showing alluvial plain facies with minor channel sandstones (espleys), and pedogenesis in the interfluvial mudstones. See p.47 for key.

the adjacent Lichfield district (Sheet 154), yielded a palynological assemblage including *Cadiospora magna*, *Mooreisporites inusitatus*, *Torispora securis* and *Thymospora* spp. (Turner, 1994) typical of Assemblage XI described by Smith and Butterworth (1967), and similar to the OT Biozone described by Clayton et al. (1977). A similar Westphalian D assemblage was described from the Halesowen Formation at Daw Mill Colliery (Smith and Butterworth, 1967, p.56), located in the adjacent Coventry district.

South Staffordshire Coalfield

In the west of the district the Halesowen Formation (Kay, 1913) crops out along the southern margins of the South Staffordshire Coalfield and in a fault-bounded sliver east of the Great Barr Fault adjacent to the Silurian Inlier at Walsall [039 970]. The formation is also present in the south-west of the district where it rests unconformably upon the Lickey Quartzite [995 782]. Most of the outcrops are disrupted by faults so that there is no continuous section from base to top of the formation.

Outcrop and borehole data from the area to the west of the district (Glover, 1990a; Powell,1991a) indicates that the formation is about 100 to 120 m thick, and broadly comprises three thick sandstone bodies, each overlain by variable thicknesses of mainly grey mudstone with thin coals and *Spirorbis* limestones. Much of the Halesowen Formation has been proved in Daleswood Farm Borehole [SO 9512 7913] (Figures 16, 18) just to the south-west of the district (Glover and Powell, 1996). Provings of the formation in the area east of the Great Barr Fault indicate that the succession here is much thinner; the maximum thickness is about 65 m, proved at Victoria (Ling) Pits [0015 9054]. In this district, the Halesowen Formation and overlying lower parts of the Salop Formation (formerly Keele Formation) are difficult to distinguish due to the common presence of primary and secondary reddening (see below).

In the type area, the Halesowen Formation rests unconformably upon the Etruria Formation. At outcrop the unconformity is difficult to detect, but on a regional scale the Etruria Formation thins rapidly westwards towards Wassell Grove Colliery (Whitehead and Eastwood, 1927, fig. 3), and eastward, across the Russell's Hall Fault (see above).

Locally, the base of the formation consists of conglomerate formed by reworking of the Etruria Formation, and clasts include igneous material, quartzite (similar to Lickey Quartzite) and vein quartz pebbles (e.g. formerly seen at Bromsgrove Street [SO 836 971]; Whitehead and Eastwood, 1927). The remainder of this lowermost sandstone consists of ripple cross-laminated and planar laminated, fine-grained sandstone with lenses rich in well-rounded caliche clasts and siliceous rhizoliths which line the base of channels. Locally, lenses of grey mudstone are present in this sandstone; indeed all three sandstones of the Halesowen Formation show considerable lateral variation in composition (Glover, 1990a). Just to the west of the district, this basal sandstone also yielded a possible conifer species (*Dadoxylon kayi*) in the form of a petrified trunk and leafy twigs of *Walchia* sp. (Newell Arber *in* Kay, 1913). The overlying mudstone succession between this basal sandstone and the second sandstone of the Halesowen Formation contains coals which were sufficiently thick to have been worked in shallow pits in Halesowen town [963 838]. Several other thin and laterally impersistent coals occur in this part of the sequence; these are generally bright coals with a well-developed partly mineralised cleat. A laterally impersistent *Spirorbis* limestone, about 0.3 m thick, also occurs at this level. The second sandstone is well exposed at Pottery's Farm [976 847], east of the Russell's Hall Fault; this medium-grained sandstone exhibits trough cross-bedding and multiple erosion surfaces. A series of grey mudstones, thin lenticular sandstones and laterally impersistent coals and limestones occur above this second sandstone. In addition, a *Spirorbis* limestone is also present; it is about 0.12 m thick and was previously referred to as the Illey Brook Limestone (Gibson, 1901). Unlike the limestones exposed in the Halesowen Formation of the Warwickshire Coalfield, this limestone can only be traced laterally for a short distance (about

1500 m). The third sandstone of the Halesowen Formation is exposed to the south and east of the type area. In the south, at Uffmoor Wood [SO 952 808], it is composed mainly of fine-grained sandstone and siltstone; this is replaced eastward by medium- and coarse-grained, cross-bedded sandstone at Illey House Farm [979 813].

The remainder of the formation comprises greenish grey and blue-grey mudstone and clayey mudstone. These are poorly exposed along the banks of streams, south of Halesowen town, and have been cored in the Daleswood Farm Borehole (Figures 16, 18) and proved in Hunnington Borehole (Figure 16) where they interdigitate with lesser amounts of red laminated and massive mudstone. A thin inferior coal, locally termed the Uffmoor Coal, occurs close to the top of the formation (Powell 1991a; Glover, 1990a); this appears to be traceable over several kilometres.

The outcrops and borehole section which have proved the Halesowen Formation east the Great Barr Fault are less informative and, in some instances equivocal in their stratigraphical classification. The sections proved at Hamstead Colliery No. 1 between 367.2 and 443.8 m depth, and at Victoria (Ling) Colliery between 14.0 and 68.7 m or 78.8 m depth and shown in Sandwell Park No.1 and Moat Farm boreholes (Figure 16) are most likely to be of the Halesowen Formation, although the top and base of the formation in the latter are difficult to pinpoint. Nevertheless, the implications are that the formation is considerably thinner (54 to 82 m) in this part of the South Staffordshire Coalfield than in the type area; this may be, at least in part, due to a lateral facies change at the top of the formation from grey beds to red beds (Figure 16; Waters, 1991a). This lateral facies change may also explain why in Heath Pits Borehole [0071 9117] and at Bullock Farm Borehole [0033 9024] the formation is also relatively thin (90 and 80 m, respectively), since the base of the Salop Formation has been drawn at the base of the predominantly red-bed sequence.

Warwickshire Coalfield

The upper part of the Halesowen Formation crops out along a narrow tract [22 96], near Kingsbury, in the north-east of the district, adjacent to the Western Boundary Fault. The formation rests unconformably on the Etruria Formation in the district, but regional evidence shows that it oversteps onto older strata. In the Dost Hill area, a few kilometres north of the district, it oversteps onto the Coal Measures over a short distance (Taylor and Rushton, 1971), and in Trickley Lodge Borehole, located about 5 km to the west of that locality, and west of the Western Boundary Fault, it rests unconformably on Lower Palaeozoic rocks. The complete succession has been proved in boreholes along the western margin of the Warwickshire Coalfield (Figure 17; Chapter Ten) where it ranges in thickness from 50 to 110 m, but on average it is about 100 m thick over most of the district.

Three cyclic (or rhythmic) units can be traced on geophysical logs (Figure 17). These upward-fining units comprise sandstone (20 to 30 m thick) at the base,

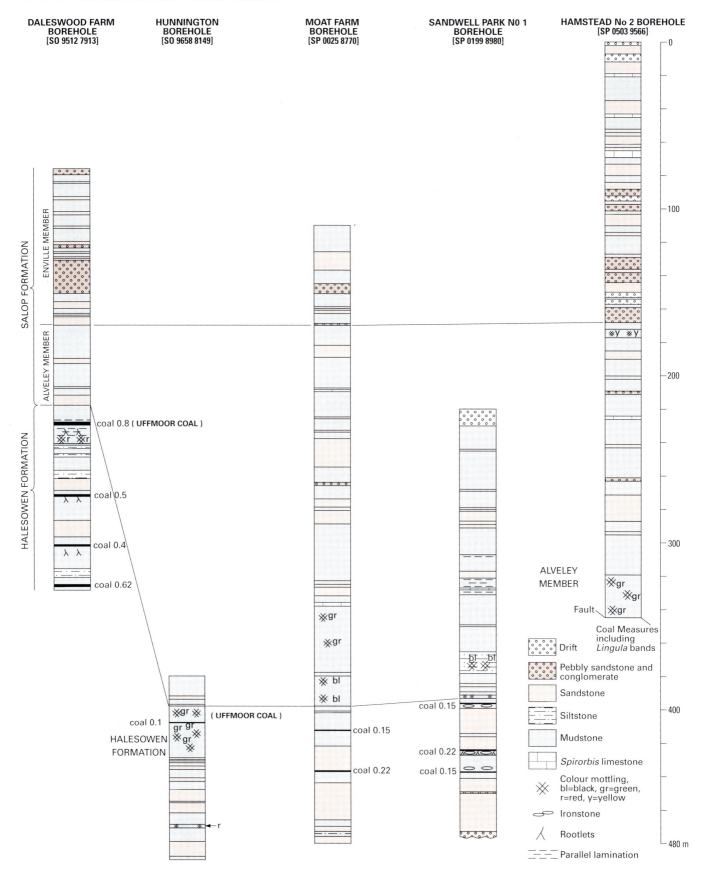

Figure 16 Correlation of the upper part of the Halesowen Formation and the lower part of the Salop Formation in the west of the district.

passing up into fine-grained sandstone and mudstone associated with thin coals and *Spirorbis* limestones. These cycles (numbered 1 to 3 in upward sequence), can be correlated with similar units in the South Staffordshire Coalfield (Glover and Powell, 1996). The Priory Wood Borehole [2361 8578] (Figure 17) proved a typical succession; the three sandstone units consist of pale grey, micaceous fine- to medium-grained, cross-bedded sandstone. The fine-grained units comprise grey and greenish grey mudstone, locally with minor amounts of finely comminuted plant debris. They are generally structureless as a result of pedoturbation; siltstones are generally ripple cross-laminated. Red and mauve colour mottling is common in the upper part of the unit, indicating early diagenetic oxidation in well-drained conditions similar to those of the overlying Meriden Formation. The formation was also proved in the Kingsbury [2446 9449] and Flanders Hall boreholes [2311 9429], in the north-east of the district, but the lithostratigraphical boundaries in these boreholes are poorly defined.

In addition to the caliche palaeosols noted above, other pedogenic features include gleysols with a carbonaceous top, semi-gleys with orange and red mottling and sphaerosiderite nodules, polygleysols with relict haematite nodules, and variable degrees of brunification (reddening). Bright vitrain-rich coals, less than 1 m thick, and commonly only a few centimetres thick are generally found capping sandstone beds.

Two thick and laterally persistent *Spirorbis* limestone beds, about 0.6 m thick, have been named. The lower bed, termed the Index Limestone, lies about 76 m above the base of the formation and can be traced throughout the Warwickshire Coalfield. It is exposed in former workings in Kingsbury Wood [229 973], located immediately north of the district. The upper bed, the Flanders Hall Limestone, lies about 107 m above the base of the formation, but cannot be traced over the whole outcrop; it was formerly worked in shallow pits [2291 9492; 2290 9550] north-west of Flanders Hall. Two *Spirorbis* limestone beds which may be equivalent to the above-named beds were proved in Kingsbury Borehole.

The upper boundary of the formation is gradational and diachronous over an interval of about 25 m; it occurs within a mudstone-dominated unit and is arbitrarily taken at the colour change from grey-green beds to red-orange coloured beds of the overlying Meriden Formation (Whitacre Member). This mudstone-dominated unit includes a gamma-ray marker (probably a coal seam) which can be traced on geophysical logs over the most of the coalfield (Bridge et al., 1998). Locally, the upper formational boundary is taken at the base of a laterally impersistent sandstone; the sandstone is poorly exposed in a quarry [2303 9484] near Flanders Hall where it exhibits lithological characteristics of both the Halesowen Formation (mica, grey-green colour) and the overlying Meriden Formation (local reddening, feldspar grains). Where it is traced northwards onto the adjacent sheet, the sandstone is lithologically similar to sandstone typical of the Meriden Formation (Whitacre Member) and is therefore included in the latter.

Depositional environments and sedimentology

The Halesowen Formation, in its type area, is characterised by three upward-fining cycles in which three laterally traceable, multi-stacked sandbodies form the base of each cycle (Glover and Powell, 1996). These three cyclic units can be traced into the Warwickshire Coalfield (Bridge et al., 1998) and the North Staffordshire Coalfield (Rees and Wilson, 1998). These sedimentary units are interpreted as having formed upon an alluvial plain — each representing a sequential change from fluvial to floodplain-dominated sedimentation developed in response to changes in sediment flux and perhaps climate. The sandstone-dominated, lower part of each cycle represents a period of local accommodation in the basin and relatively high flux; this resulted in the lateral and vertical stacking of channels, due to the lower preservation potential of floodplain fines. The presence of thin coals and *Spirorbis* limestones associated with the upper, fine-grained part of the cycles suggests that, at times, the water table was high enough and sediment input sufficiently low to allow the local, coeval development of peat (in rheotrophic swamps) and lacustrine carbonates. In contrast, the uncommon preservation of caliches (in situ and reworked) indicates periods of relatively high rates of evapotranspiration. Furthermore, the local presence of brunification and ferruginous palaeosols, particularly in the upper part of the formation, suggests that better-drained conditions developed locally on the alluvial floodplain. Penecontemporaneous breccia textures and rhizolith tubules in the *Spirorbis* limestones suggest fluctuating lake levels and periodic emergence.

The heavy mineral suite of the Halesowen Formation sandstones indicates that sediment was derived from a hinterland comprising low- to moderate-grade metasediments, probable granites and some ultramafics. The Pennant Sandstone (Bolsovian) of the South Wales Coalfield is petrographically similar and this is known to have been derived from the south (Kelling, 1974). Hallsworth (1992a) and Glover and Powell (1996) suggested that the Halesowen Formation source rocks were likely to have been in the Variscan hinterland; the presence of chloritoid indicates that some of the debris may have been derived from the Île de Groix in Brittany.

SALOP FORMATION (SOUTH STAFFORDSHIRE COALFIELD)

The term Salop Formation supersedes the terms Keele and Enville formations (Besly and Cleal, 1997) and refers to the remainder of strata assigned to the Upper Carboniferous in the western part of the district. These predominantly red-bed strata had formerly been subdivided into a mudstone-dominated succession, the Keele Beds (Whitehead and Eastwood, 1927) or Keele Formation (Besly, 1988), and a sandstone-rich upper succession known variously as the Enville Group (Hamblin, 1982), the Enville Formation (Besly, 1988), the Calcareous Conglomerate Group of the Enville Beds (Whitehead and Eastwood, 1927; King, 1899; King, 1923) or the Hamstead

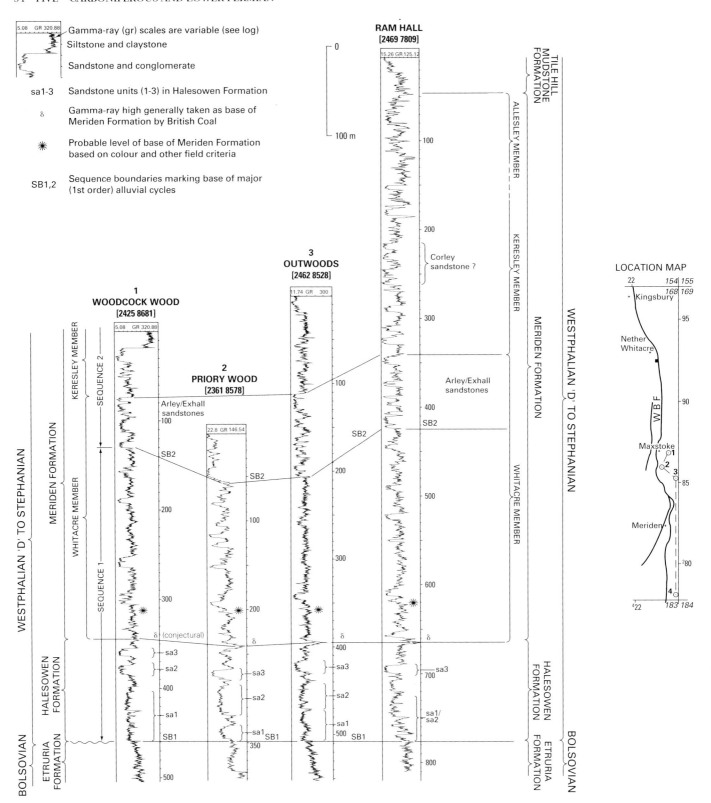

Figure 17 Correlation of geophysical logs of the upper Carboniferous succession along the south-west margin of the Warwickshire Coalfield. Based on British Coal (Coal Authority) data. WBF Western Boundary Fault.

Beds (Boulton, 1924). The sandstones of the upper part of the succession (Enville) are different from those of the lower mudstone-dominated unit (e.g. Whitehead and Eastwood, 1927, p.133) in that they contain pebbles of red and yellow chert. Limestone pebbles, both Silurian and Carboniferous, are also abundant at some localities, as well as pebbles of Llandovery sandstone and quartzite and vein quartz of unknown affinity. The distinction between the sandstone- and mudstone-rich units is also retained in this new classification; the lower mudstone-dominated unit is termed the Alveley Member and the sandstone-rich unit is named the Enville Member. The members are identified in boreholes, but have not been shown on the 1:50 000 scale map because of the difficulty in identifying the sequence in faulted or drift-covered ground. The Salop Formation is lithostratigraphically equivalent to the lower part of the Meriden Formation of the Warwickshire Coalfield, in the east of the district.

The Salop Formation transitionally overlies the Halesowen Formation and crops out in the south-west part of the district and to the east of the Great Barr Fault. The formation varies considerably in thickness; in the south it is around 110 m thick, where it is unconformably overlain by the Clent Formation, and to the east of the Great Barr Fault it may be as much as 500 m thick.

The **Alveley Member** (Besly and Cleal, 1997) is present in Daleswood Farm Borehole [SO 9511 7913] where it consists of 49 m of mainly massive red mudstone (Figure 18), with less common laminated beds as well as several caliches, especially towards the top of the member, and thin sandstone beds. The equivalent beds crop out in the southern part of Uffmoor Wood [SO 957 809]; in this area a sandstone, laterally traceable for over 500 m locally marks the base of the Salop Formation. Sandstones in this part of the Salop Formation are commonly fine to medium grained and contain beds and lenses of red angular intraformational mudstone clasts. Two *Spirorbis* limestone beds are present in this lower part of the Salop Formation at Bartley Reservoir [005 813]. The details of the geology of the area now beneath the reservoir was recorded mainly by Boulton (1924, 1933) with sections from Barnes (1927). This part of the Salop Formation contains a high proportion of sandstones compared with the mudstone-rich succession seen elsewhere in the South Staffordshire Coalfield. Recent site investigations have also proved a *Spirorbis* limestone at the nearby Frankley Reservoir [001 801] (Glover, 1990b).

To the east of the Great Barr Fault, the mudstone-dominated part of the Salop Formation forms much of the Upper Carboniferous outcrop. The base of the Salop Formation in this part of the South Staffordshire Coalfield has commonly been drawn at the base of a thick sandstone (Waters, 1991a). The overlying mudstone-dominated succession locally includes a thin coal and associated fireclays, and thin and laterally impersistent sandstones (Waters, 1991a). *Spirorbis* limestones have been proved in several boreholes in the South Staffordshire Coalfield near the top this mudstone-dominated part of the Salop Formation (Whitehead and Eastwood, 1927, pl. VI). These thin lacustrine limestones are not correlatable across the coalfield, but it is possible that they occur at broadly similar stratigraphical levels forming localised lenses.

The **Enville Member** of the Salop Formation has also been cored in the Daleswood Farm Borehole [9511 7913] which proved about 97 m of multistorey, erosively based red-brown medium- to coarse-grained, trough cross-bedded sandstones, rich in intraformational clasts of caliche and red mudstone and, less commonly, extraformational limestone (Figure 18). Intercalated mudstones contain well-developed caliche palaeosols. The topmost part of the Salop Formation has been cored in the Romsley Borehole [SO 9501 7893] (Figure 18) which proved sandstones, with erosional bases, and red mudstone with caliche beneath the Clent Formation. East of the Great Barr Fault, the sandstone-rich part of the Salop Formation outcrops at Bristnall Fields [9981 8693] and Sandwell Park Golf Course [0226 9100]. Outcrops of sandstones containing amphibian or reptilian footprints (Haubold and Sarjeant, 1973) as well as fossil plants (*Walchia* sp. and cordaites) (Hardaker, 1912) are no longer exposed. Nearby, a canal section at Tower Hill [0525 9291] exposes the sandstone-rich Enville Member comparable with that proved in the Hamstead No. 2 Borehole (Figure 16).

Extraformational conglomerates containing subrounded and well-rounded pebbles of Carboniferous limestone, dolomite, silicified limestone, quartzite and sandstone are exposed at Barnford Hill Park [997 875]. Similar extraformational conglomerates, exposed in a railway cutting [0251 8977 to 0291 8956] near Watville Bridge (King, 1923; Powell, 1991c) yielded oolitic, bioclastic limestone and chert clasts (Carboniferous Limestone), including a proetid trilobite cf. *Archegonus* (Rushton, 1990). In the north of this area, the Salop Formation appears to be mudstone-dominated with only a few lenticular beds of sandstone and calcareous conglomerate present; this may reflect a gradual northwards increase in mudstone beds in the upper part of the Salop Formation.

Angular clasts of fine-grained, possibly tuffaceous, material have been described from the sandstones in the upper part of the Salop Formation formerly exposed east of Wildnerness Lane [041 948] and exposed in the stream section at Forge Farm [0226 9255 to 0238 9263]. On the basis of clast type these beds may be transitional between the Salop and the overlying Clent Formation.

MERIDEN FORMATION (WARWICKSHIRE COALFIELD)

In this area, the succession above the Halesowen Formation predominantly comprises red-bed, alluvial strata, generally devoid of regional lithostratigraphical and biostratigraphical markers. The lithostratigraphical nomenclature of the sequence in the Warwickshire area was based on early 19th century studies of coeval sequences in North and South Staffordshire (King, 1899; Gibson, 1901; Newell Arber, 1916). Two broad divisions were recognised, namely the Keele 'unit', largely consisting of sandstone and mudstone, and the overlying Enville 'unit', of similar lithology, but with a higher proportion of sandstone, and including pebble-conglomerates rich in extrabasinal clasts. The nomenclature of these broad subdivisions was later adapted by subsequent workers (Shotton, 1929; Old et al., 1987, 1990; Besly, 1988), largely on the basis of the proportion of sandstone to mudstone (Table 13). Subsequently, a new lithostratigraphical nomenclature (Table 13), based on both field characteristics and geophysical wireline log signatures in coal exploration boreholes, was introduced for the Warwickshire area in the Coventry Memoir (Bridge et al., 1998), to which the reader is referred for details of the scheme. In outline, the Keele Formation and Coventry Sandstone Formation (Enville Group) of Old et al., (1987; 1990) are replaced by the Meriden Formation which is subdivided into three members (see below). The revised scheme (Table 13) differs from that used, herein, for coeval strata in the South Staffordshire area (Glover and Powell, 1996), where the lithostratigraphical units have different characteristics and stacking patterns, probably as a result of variation in sediment flux, provenance and depositional settings in the two areas.

The Meriden Formation crops out along the western margin of the Warwickshire Coalfield and has been proved in many boreholes (Figures 17, 21). It consists of red and red-brown mudstone, and sandstone with subordinate conglomerates, breccias, and thin *Spirorbis* limestones. The formation reaches a maximum thickness of about 675 m in the Coventry district (Bridge et al., 1998). The age of the lower part of the formation is Westphalian D; this is based on a nonmarine bivalve fauna (Old et al., 1987), and correlation with coeval beds in South Staffordshire which include thin coals that yielded a monolete miospore assemblage indicative of a Westphalian D age (Owens, 1990). The upper part of the formation is probably Stephanian age; the biostratigraphical evidence is both sparse and somewhat equivocal, (see Poole, 1975 and discussion in Waters et al., 1995, and references therein). However, a jaw bone of the pelycosaur *Ophiacodon* sp. collected from the upper part of the formation in the Coventry area (Murchison and Strickland, 1840) was assigned a late Stephanian to Early Permian age (Paton, 1974).

The Meriden Formation, as defined by Bridge et al. (1998) is subdivided, in ascending sequence, into the Whitacre Member, Keresley Member and Allesley Member. These units, interpreted as upward-coarsening cycles (Bridge et al., 1998), are described below.

Whitacre Member

The Whitacre Member (Figure 17) is equivalent to both the former Keele Formation and the lowermost sandy,

Figure 18   Graphic logs of the Daleswood and Romsley boreholes showing facies associations in the Halesowen, Salop and Clent formations (after Glover and Powell, 1996).

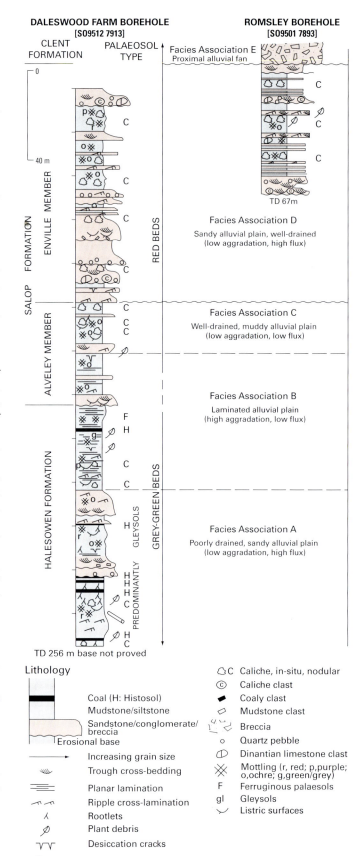

pebbly units of the Coventry Sandstone (Enville Group) of Old et al. (1987, 1990). In the district, the member is about 270 m thick, but geophysical logs in the Coventry district indicate a maximum thickness of 325 m (Bridge et al., 1998). Red mudstone and subordinate, thin beds of fine- to medium-grained sandstone comprise the lower part of this member, but the unit becomes increasingly sandy towards the top; the uppermost (60 to 130 m) sandstone-dominated strata are informally termed the Arley and Exhall sandstones (Bridge et al., 1998), which include beds rich in extrabasinal pebbles of Carboniferous limestone and chert (Eastwood et al., 1925; Shotton, 1927).

The sandstones are litharenites, composed of mono-crystalline quartz, rock fragments, including non-ferroan micrite (caliche and Dinantian limestone), chert, polycrys-talline quartz/chert, cryptocrystalline silica, sparse leached sodic and potassic feldspars, and sparse mica (Lott, 1992b). Haematite grain coatings are common, and cements comprise discontinuous quartz overgrowths with patchy non-ferroan spar, sparse slightly ferroan spar and patchy kaolinite pore fill (Lott, 1992b). Intraformational red mudstone rip-up clasts (up to 0.3 m diameter), reworked caliche nodules and, less commonly, fragments of silicified roots (rhizoliths) infill erosive scours in channel sandstones. Rounded and subrounded red (haematite) and brown (goethite) clasts are common in the sandstones in the lower part of the member; some of these clasts show concentric lamination and are inter-preted as reworked ferruginous palaeosols. Bedforms generally comprise planar and trough cross-bedding in laterally wedging cosets. The sandbodies are laterally impersistent, and, traced across country, they split and amalgamate over distances of a few hundred metres. Quarries [241 880; 2415 8850; 2416 8806] in the Arley sandstone, near Hill Farm, Maxstoke (Figure 19), reveal pebbly sandstones and conglomerate lenses (up to 0.5 m thick) which include extrabasinal clasts including grey, shelly Carboniferous limestone, chert, and purple sand-stone, together with vein-quartz and porphyry (Eastwood et al., 1923, p.82; Shotton, 1927). Interbedded red and purple mudstone, which comprises most of the member, is generally structureless, probably as a result of pedoturba-tion; green reduction haloes and desiccation cracks are locally common, and pedogenic features include common caliche nodules and pervasive brunification.

Thin beds of *Spirorbis* limestone, consisting of grey, micritic limestone, platy or nodular in form, and charac-terised by clotted peloidal, and brecciated textures, are present throughout the unit. Three of the more persis-tent beds have been named (Eastwood et al., 1923) and have been traced as marker beds over most of the coalfield; despite their local use in correlation, the resurvey has shown that the beds cannot be traced over the whole district. The named beds, listed in ascending order, together with their approximate height above the base of the Whitacre Member are Baxterley Limestone (40 m), Whitacre Hall Limestone (90 m) and Maxstoke Limestone (250 m). The *Spirorbis* limestones generally form irregular, platy fragments in the soil and can be traced across country on the evidence of the abundant

soil brash and the presence of numerous shallow pits dug along their outcrop. An isolated exposure in a stream bed near Hermitage Farm [2381 8575] revealed 0.15 m of limestone which yielded ostracods but no *Spirorbis* (Cantrill, 1909). A temporary exposure [2399 9482 to 2399 9489] in the lower part of the member showed a limestone composed of small lens-like clusters (0.2 m high by 0.4 m long) of grey, nodular and platy micrite which pass laterally into red mudstone; the lens-like limestone clusters are distributed at irregular intervals along a bedding plane. Thin sections revealed a clotted, peloidal micrite with rhizolith tubules, but no *Spirorbis*. This limestone is interpreted as early diagenetic, pedogenic carbonate, but the thicker, laterally persistent, named beds which contain *Spirorbis* and sparse ostracods were probably deposited in ponded, shallow lakes.

Keresley Member

The Keresley Member, as defined by Bridge et al. (1998) crops out in the south-west of the coalfield, and is about 270 m thick in this district. The lower part is mudstone-dominated and the proportion of sandstone increases gradationally upward; the upper 85 m or so consist of thickly bedded sandstone with minor, thin beds of mud-stone and conglomerate, informally known as the Corley sandstone. Beds of *Spirorbis* limestone are also present but, in contrast to the underlying Whitacre Member, none is sufficiently persistent to be used as regional marker beds.

Lithologies in the Keresley Member are similar to the Whitacre Member (see above for details). Pebble-rich sandstones and conglomerates, including a large propor-tion of Silurian clasts, typical of the upper part of the member (Corley sandstone) in the Coventry district (Shotton, 1927; Bridge et al., 1998) have not been recog-nised here. In this district, the upper part of the member comprises beds of fine- to coarse-grained sandstone, up to about 10 m thick, intercalated with mudstone (Ram Hall Borehole; Figure 17). Intraformational red mud-stone rip-up clasts commonly line erosive scours in the sandstones. Coeval strata in the Coventry district yield a diverse conifer flora, including *Lebachia* spp. (Vernon, 1912; Eastwood et al., 1923; Wagner, 1983).

Allesley Member

This member forms the highest subdivision of the Meriden Formation, and crops out in the south-east corner of the district where it is about 125 m thick. It consists of fine- to coarse-grained sandstone interbedded with red mudstone, lithologically similar to the underlying members, although in the type area to the east of the district (Bridge et al., 1998), the lower part is mudstone-dominated (Ram Hall Borehole, Figure 17).

A sandstone bed with a conglomerate lens, which crops out near Berryfields Farm [246 814; 248 811], contains abundant fragments of black, red and cream sili-cified wood (Sumbler, 1982) together with clasts of Carboniferous limestone and *Spirorbis* limestone. This bed is probably a correlative of the Allesley conglomerate which crops out about 5 km to the east, and which yielded fossil wood identified as *Cordaites brandlingi* (Eastwood et al., 1923, p.27).

DEPOSITIONAL ENVIRONMENTS AND SEDIMENTOLOGY OF THE MERIDEN AND SALOP FORMATIONS

The lithological and sedimentological characteristics of the red-bed succession overlying the Halesowen Formation indicate deposition in both proximal and distal alluvial plain settings, in a semi-arid climatic regime (Wills, 1956; Besly, 1987; Besly, 1988; Glover and Powell, 1996). Red mudstones and associated brunified alluvial palaeosols and ferruginous palaeosols were deposited in well-drained (oxidising) conditions, on the interfluves, during phases of floodplain aggradation; sparse laminated mudstones suggest deposition in shallow lakes, but pervasive pedogenesis probably destroyed much of the original lamination. High rates of evapotranspiration on the alluvial plain resulted in the formation of nodular caliches, which were locally reworked as pebble-sized clasts and incorporated into sands during flood events. *Spirorbis* limestones were deposited in broad, shallow lakes fringed by plants (rhizolith tubules); some of the lakes must have extended over many kilometres, judging from the widespread occurrence of the named limestones in the Whitacre Member. Siliciclastics, comprising siltstone, sandstone and locally conglomerate, represent increased sediment flux and deposition in ephemeral low-sinuosity channel, sheetflood and proximal alluvial fan settings.

In the Warwickshire area, Besly (1988) considered that the formation was deposited as both proximal (sandstone and conglomerate) and distal (mudstone) alluvial fans. This concept was later adopted by Bridge et al. (1998) who subdivided the sequence based on the recognition, both in the field and on wireline logs, of broad, upward coarsening trends (the subdivisions of the Meriden Formation). However, coeval strata in the South Staffordshire area (see previous section), together with the underlying Halesowen Formation, have been interpreted as the product of upward-fining alluvial sequences, controlled to a great extent by variation in sediment flux, which in turn was influenced by both climate and tectonics (Glover and Powell, 1996). The difference in interpretation focuses on the relative importance placed on mudstone-dominated units as the base of cycles (members of Bridge et al., 1998), compared to the influx of coarse-grained siliciclastics associated with conglomerates rich in exotic clasts (Enville Member of South Staffordshire, and the Arley/Exhall sandstones and Corley sandstone of Warwickshire). Geophysical log signatures, particularly the gamma-ray log (Figure 17), can be interpreted as upward-fining alluvial cycles, about 150

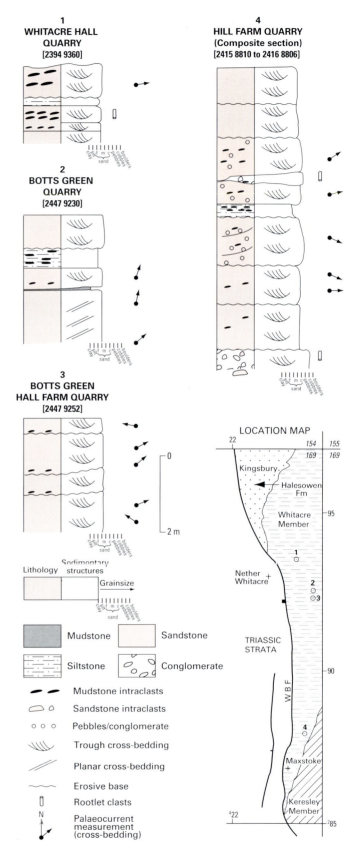

Figure 19   Graphic logs of sandstone/conglomerate beds in the Whitacre Member (Meriden Formation) showing principal palaeocurrent directions.

to 300 m thick (Glover and Powell, 1996), the base of which represents a sequence boundary (SB on Figure 17). The sandstone (and locally conglomerate) which form the lower part of the cycles, are interpreted as the deposits of alluvial channels; these pass upwards to finer-grained sandstones and mudstones, deposited during floodplain aggradation. The major sequences (cycles S1 and S2) are compared to the lithostratigraphical scheme in Figure 17.

The Whitacre Member was thought to have a northerly provenance (Besly, 1988), possibly derived from a region of low-grade metasediments. However, the sparse palaeocurrent data in the district (Figure 19) and the Coventry district (Bridge et al., 1998) indicate a northward or north-eastward palaeoflow. Study of the larger clasts in the conglomeratic sandstones (Arley/Exhall sandstone and Corley sandstone) led Shotton (1927) to propose a derivation from the west for the former, and a derivation from the east for the latter. The source areas were believed to be uplifted horst blocks of Lower Palaeozoic strata which occupied the sites of the present-day Triassic Knowle and Hinckley basins, located to the west and east of the Warwickshire coalfield, respectively. However, study of similar clasts, in coeval strata of South Staffordshire suggests a Lower Palaeozoic component derived from the east and south, possibly from an uplifted part of the present-day Worcester Basin and Neoproterozoic Uriconian rocks (see South Staffordshire section).

A potential source area is indicated on the pre-Devonian subcrop map (Smith 1987, fig. 1) extending from the Worcester Graben into the Knowle Graben. This is now known to be subdivided into two subcrop areas of 'Precambrian' and Lower Palaeozoic rocks (Chadwick and Smith, 1988).

In summary, there appears to be good evidence for derivation from mixed source areas located around the southern, south-eastern and south-western margins of the Pennine Basin. Coarse-grained sediment flux (second-order cycles) may have been controlled by tectonics (geomorphic uplift and movement along reverse faults) associated with Variscan deformation of the Wales-Brabant Massif (Glover and Powell, 1996).

CLENT FORMATION (SOUTH STAFFORDSHIRE)

The Clent Formation is equivalent to the Clent Breccia or Breccia Group (King, 1899, 1923b; Eastwood et al., 1925; Whitehead and Eastwood 1927); the term breccia has been dropped since, in many of the northerly exposures of this formation, breccia is subordinate to mudstones and sandstones (Glover, 1991a; Powell, 1991a). The Clent Formation crops out in the south-west of the district forming the hills of Romsley and Dayhouse Bank, and Windmill Hill, to the south-east. There are no exposures of the formation in these areas; the assumption that the Clent Formation crops out in these areas, along with four other outcrops in the south-western part of the district, is made on the basis of a distinctive soil type, rich in angular to subangular volcanic

pebbles. These and other extraformational clasts are typical of the Clent Formation in its type area, located to the west (Powell, 1991a). Exposures are rare because the muddy matrix of the breccia disintegrates at the surface. The only exposure of the Clent Formation in the district is present in a road cutting towards the bottom of Merritt's Hill Lane in Northfield [015 805]. This section exposes almost 10 m of thick and, in places, lenticular, structureless beds of pebble conglomerate, rich in angular and subangular pebbles, supported in a matrix of red mudstone. Clasts include 80 to 84% of igneous rocks (mainly fine-grained tuff), 14 to 16% Llandovery limestones, and 2 to 4% calcareous sandstones, with minor amounts of chert (Boulton, 1933). Approximately 8 m of matrix-supported breccia have also been proved in the Romsley Borehole, which lies close to the south-western corner of the sheet (Figure 18).

The relationship between the Clent Formation and the underlying Upper Carboniferous strata has been the centre of debate since the turn of the century. Boulton (1924, 1933) concluded that the Clent Formation unconformably overlies the Upper Carboniferous strata in the type area (Whitehead and Pocock, 1947, p.94); this relationship is substantiated by the presence of the Clent Formation resting directly upon the lower, mudstone-rich part of the Salop Formation at Merritt's Hill. The presence of a thin conglomerate with clasts similar to those present in Enville Member (formerly Enville Formation) sandstones (Fleet, 1927) at the base of the Clent Formation, as proved in a well at Merritt's Hill, was argued by Boulton (1933) to represent reworking of Enville sandstones during deposition of the Clent Formation.

Whether a major unconformity exists beneath the Clent Formation is still largely open to debate (King, 1923b; Whitehead and Pocock, 1947). The gradational boundary between the mudstone/sandstone-dominated Salop and Clent formations in distal basinal areas farther to the north-west, in the Wombourne and Penn districts (Glover, 1991b; Powell, 1991c), suggests that the unconformity (or disconformity) is less marked here than in the Clent Hills, located close to the southern margin of the basin. In distal locations the boundary does not represent a significant tectonic break in the succession. For these reasons the Clent Formation is still best regarded as being genetically related to the underlying depositional succession than to younger Permo-Triassic strata such as the Hopwas Breccia (see Chapter Six). The rapid lateral lithofacies change from breccias and conglomerates in the south, to mudstones and sandstones in the north, may indicate that some of the mudstone-rich beds proved in boreholes below the Triassic succession between the Birmingham and Great Barr Faults may be equivalent to the Clent Formation rather than the Salop formation.

Clent Formation proved in boreholes in central Birmingham

Boreholes sunk for water on the footwall side (north-west) of the Birmingham Fault in central Birmingham (e.g.

Nechells Gasworks Nos. 1 and 2, Windsor Street Gasworks No. 3 and Dunlop Rubber) proved a succession of breccia and sandstone beds, termed the Nechells Breccia (Boulton, 1924, 1928, 1933; Butler and Lee, 1943). These beds overlie the Salop Formation and are overlain by Triassic Hopwas (Quartzite) Breccia or, where this is absent, by Kidderminster Formation conglomerate.

Boulton showed that the clast and sandstone petrography of these breccia/sandstone beds in the central Birmingham boreholes is locally variable but, in general, they are similar to those at crop at Warley and Northfield, in west Birmingham, and to the Clent (Breccia) Formation of the type area (Clent Hills), located to the south-west of the district (Powell, 1991a). Consequently, the term Clent Formation (Besly, 1988) is used here for all these geographically isolated breccia beds which are considered to be approximately coeval.

In the Nechells Gas Works No. 1 Borehole and the Windsor Street Gas Works No. 3 Borehole, the Clent Formation is 92 m and 113 m thick, respectively. In the Nechells Borehole the formation consists of beds of fine to very coarse breccia (up to 13 m thick) interbedded with beds of brown, red, and less commonly, grey, calcareous and non-calcareous sandstone and marly sandstone containing angular fragments and pebbles. The breccia and sandstone beds have strongly erosive bases. Clasts in the breccia are angular to subangular and range in size from 0.15 m to 0.22 m long axis. Together with the smaller clasts in the sandstone beds, they predominantly comprise igneous rocks, including volcanic, crystal-rich tuff, rhyolite, felsite, andesite and basalt. Aeolian features such as wind polishing, haematite patina and etching typical of dreikanters were noted on some of the clasts by Boulton (1933). The igneous clasts were considered by Boulton (1924, 1933) to be derived from the Proterozoic Uriconian rocks of Shropshire. Cambrian sandstone and limestone clasts are also common in the Nechells Borehole; they contain a diagnostic fauna including *Paterina, Camenella, Torellella, Hyolithellus, Coleoloides, Salterella, Micromitra, Obolus, Runcinodiscus and Strenuella* (Plate 3a–l; see Chapter Three for details) (Boulton, 1924; Rushton, 1994b). Boulton (1933) noted that sandy limestone clasts in the nearby F Smith's Brewery Borehole are similar to the Cambrian Hyolithus Limestone beds of Woodlands Quarry, Atherstone, near Nuneaton. This provenance was supported in a recent study of the Nechells Borehole clasts (Rushton 1994b; Chapter Three), which showed that one suite of limestones is similar to the Home Farm Member of the Hartshill Sandstone Formation of the Nuneaton succession (Bridge et al., 1998). The limestone clasts may have been sourced from Cambrian strata located to the east, either from the Nuneaton area, or probably from a more local area such as the present-day Knowle Basin which, prior to basin subsidence in Permo-Triassic times, may have been an uplifted source area of Neoproterozoic and Cambrian rocks (see Chapter Nine). A single, large limestone clast containing a Cambrian fauna similar to that of the Comley Limestones of Shropshire was also identified (Rushton, 1994b), but the size of the clast suggests that it was derived from a more local source,

possibly the Knowle Basin area, even though no similar calcareous beds are known from the *Strenuella* Biozone of the Nuneaton succession located farther to the east. In addition, Carboniferous limestone and chert clasts, possibly secondarily derived from conglomerates in the underlying Salop Formation (see above) are present in the Windsor Street Borehole cores.

Heavy mineral residues at 179.8 m and 226.2 m depth (Boulton, 1924) showed a general scarcity of garnet, staurolite, anastase and tourmaline, and yielded a lower proportion of apatite than in the overlying Triassic Hopwas (Quartzite) Breccia,

Depositional environments and sedimentology

The bedforms and textures suggest that the Clent Formation breccias at outcrop in the district and in the Clent Hills, were deposited as a series of proximal alluvial fans which, locally incorporated clasts showing aeolian weathering. On the basis of clast lithologies and lateral facies relationships, the alluvial fans are believed to have been largely derived from Neoproterozoic (Precambrian) igneous rocks located to the south-west of the district or from an uplifted area to the east of the Birmingham Fault (e.g. beneath the Permo-Triassic succession of the Worcester Basin).

The presence, in the central Birmingham boreholes, of angular and subangular igneous clasts similar to the Uriconian igneous rocks of Shropshire (located 80 km to the west), together with Cambrian limestone/sandstone clasts, possibly derived from the area east of the Birmingham Fault, may indicate a dual provenance (Boulton, 1933). However, the angular nature and large size of the breccia clasts, and their deposition in an alluvial fan environment, suggests a source area located closer to central Birmingham than the Uriconian outcrop in Shropshire. It is proposed that these rocks were derived from uplifted 'basement' Neoproterozoic and Lower Palaeozoic rocks located, either to the south, in the present-day northern part of the Worcester Basin (Boulton, 1933; Wills, 1956; Chadwick and Smith, 1988), or an area to the east of the Birmingham fault, probably the present-day Knowle Basin.

TILE HILL MUDSTONE FORMATION

The Tile Hill Mudstone Formation (Shotton, 1929; Bridge et al., 1998) represents the uppermost late Carboniferous to early Permian red-bed unit of the Warwickshire coalfield, in this district. The lowermost part of the formation crops out in the south-east corner of the district; about 45 m of red mudstone with thin sandstone beds, assigned to this formation were proved in the Ram Hall Borehole (Figure 17). It consists of red and brown, blocky mudstone and laminated siltstone with subordinate, laterally impersistent, thin beds of red-brown, fine- to coarse-grained sandstone, locally with pebbles and conglomerate lenses. The base of the unit is marked by a thick, regionally traceable mudstone sequence (Bridge et al., 1998).

No definitive biostratigraphical age has been proved, but the superposition of the formation below the Kenilworth Sandstone Formation, which crops out in the Warwick district to the south-east, suggests a late Stephanian or early Permian (Autunian) age. The latter determination is based on the presence of the pelycosaurs *Sphenacodon brittanicus* and *Haptodus grandis* and the amphibian *Dasyceps bucklandi* in the Kenilworth Sandstone of the Warwick district (Paton, 1974, 1975; Old et al., 1987).

The Tile Hill Mudstone was deposited in alluvial plain environments, similar to those which prevailed during deposition of the Meriden Formation; however, deposition was dominated by fine-grained siliciclastic sedimentation in more distal floodplain settings.

CARBONIFEROUS IGNEOUS ROCKS

Dolerite

Carboniferous igneous rocks occur at outcrop (Figure 1), and have been proved at depth, in the South Staffordshire Coalfield. The rocks comprise alkaline olivine dolerite and basalt, which intrude the Coal Measures and Etruria Formation (Figure 7). Their age is inferred, from stratigraphical evidence, to be Bolsovian (Westphalian C), due to the absence of igneous rocks intruding strata of Westphalian D age and younger. This age is supported by correlation with volcanic rocks of undoubted Bolsovian (Westphalian C) age, associated with similar intrusions, which crop out at Tansey Green [SO 910 896], west of Dudley (Glover et al., 1993). Radiometric dating of the intrusive dolerites, using the potassium-argon method (Kirton, 1984), provides inconclusive dates as young as Permian times; these are considered to represent minimum ages.

Early workers, including Jukes (1859) and Pocock (1931), proposed an extrusive origin for some of the igneous rocks. However, Eastwood et al. (1925) and Marshall (1942; 1945) showed that the igneous rocks in the district are intrusive in origin, although undoubted extrusive volcaniclastic rocks are present in the Tansey Green area [SO 910 896], west of the district (Glover et al., 1993).

A single, large intrusive body crops out in the Rowley Regis–Darby's Hill area [96 87 to 96 89], forming a prominent north–south-trending ridge about 3 km long (Plate 1). This body, locally referred to as the 'Rowley Rag', is the largest Carboniferous basic igneous intrusion in the West Midlands and has been the subject of numerous publications (Murchison, 1839; Allport, 1870; 1884; Pocock, 1931; Marshall, 1942).

The intrusion, formerly referred to as a laccolith (Eastwood et al., 1925), is in fact a lopolith, with a planar upper intrusive contact and a markedly, downward-convex lower intrusive contact. The lopolith is about 100 m thick in the central part, where it has been extensively worked (Chapter Two) at Hailstone [966 882] and Edwin Richards quarries [968 884], and it appears to decrease rapidly in thickness toward the margins. The lopolith has been almost entirely undermined for coal (Figure 3) and no evidence of a feeder pipe has been found (Eastwood et al., 1925). It is likely that the intrusion was fed laterally through marginal sills. The geometry of the lopolith has been extensively disrupted by post-intrusive faulting (see Chapter Nine).

Extensive working of the dolerite at Hailstone and Edwin Richards quarries has revealed the presence of three distinct joint patterns. A capping of relatively massive dolerite shows widely separated joints, irregular vertical joints are present in the core of the body, and listric joints are present in the periphery of the body, occurring parallel with the igneous contact.

The petrography of the Rowley Regis dolerite has been described by Allport (1884), Eastwood et al. (1925). Thin sections, from a borehole drilled at Turner's Hill [967 886], show the following petrography:

Thin section E62881 (collected near to top of intrusion) shows a porphyritic texture with altered olivine phenocrysts (1.0 to 1.5 mm diameter) and subordinate augite phenocrysts (0.75 mm diameter) that commonly form glomerophytic masses. The groundmass displays an ophitic or subophitic texture with laths of plagioclase overgrowing small clinopyroxene crystals. Euhedral ilmenite crystals are a common constituent of the groundmass.

Thin section E62882 (collected from central part of intrusion) is similar to specimen E62881, but is more porphyritic with phenocrysts of olivine and plagioclase (1.0 to 2.0 mm diameter) surrounded by a plagioclase-dominated groundmass.

Thin section E62883 (collected from about 5 m above the base of the intrusion) is holocrystalline in texture, with alignment of plagioclase laths.

The basal intrusive contact of the Rowley Regis dolerite was exposed at the time of the survey in Hailstone Quarry [9665 8835]; above the contact, a 0.5 m-thick zone of dolerite, with joints trending parallel to both an igneous lamination and the intrusive contact, was observed. Above this zone, near-vertical cooling joints dominate. The Etruria Formation below the intrusive contact comprises a 1.5 m-thick sandstone ('espley'), underlain by black, hardened Etruria Formation mudstone which was thermally altered during the intrusion of the dolerite. A near-vertical contact between the dolerite and Etruria Formation mudstone, on the western flank of the lopolith, was exposed at Hailstone Quarry [9655 8842].

The top of the Rowley Regis dolerite was observed at Darby's Hill Quarry [9683 8962], which is now backfilled. Here, the base of the section comprises about 1.6 m of finely crystalline dolerite; the intrusive contact is overlain by about 1.8 m of red, mottled silty mudstone, showing no apparent effects of thermal metamorphism. Dolerite was also exposed in disused quarries at Rough Hill [9625 8895], Yewtree Lane [9675 8690] and Bury Hill Park [9770 8915].

Dolerite and basalt sills have been proved in a number of shaft records, but have not been found at outcrop in the district. A sill, about 1 m thick, was proved in a borehole at Turner's Hill [967 886] below the main lopolith at Rowley Regis. The sill consists of highly chloritised basalt with faint relicts of plagioclase phenocrysts and abundant spherical, calcite-filled amygdales, up to 2.5 mm diameter. Colliery shafts proved dolerite intruding the Coal Measures, particularly at stratigraphical levels below the Bottom Holers Coal. For example, in Bradley Colliery No. 15 Pit [9602 9485] a sill at least 25 m thick, was proved at the base of the shaft, a sill at least 3 m thick was proved at the base of Wallbutts Colliery No. 3 Pit [9608 9621], a 12.8 m dolerite body was proved in Moxley Colliery No. 5 Pit [9625 9620], and thin sills were proved at Tividale Colliery [around 962 901]. The Bottom and Bottom Holers coals, adjacent to the sills, show variable degrees of thermal alteration.

SIX

Triassic

Triassic rocks underlie most of the district and comprise predominantly continental, fluvial and lacustrine red-bed sediments which were deposited in an arid to semi-arid climatic regime at low northern palaeolatitudes. However, parts of the succession were deposited in estuarine or marine environments. The predominant red colour of the strata is due to early diagenetic oxidation of detrital ferromagnesian silicates and iron-bearing clay minerals to iron oxide (haematite) (Walker, 1976). The thick succession in the Knowle Basin (Figures 1, 29) was deposited in an extensional (or transtensional) tectonic regime (Chadwick and Smith, 1988).

The terrigenous Triassic strata of Britain were formerly divided into the 'Bunter' and the overlying 'Keuper', names adopted from the Triassic of Germany (Sedgwick, 1829) for the predominantly arenaceous and argillaceous strata, respectively. However, Warrington et al. (1980) formally revised the lithostratigraphical nomenclature (Table 14), and their scheme is adopted herein. The Hopwas Breccia, present locally at the base of the

Table 14 Stratigraphical nomenclature of the Triassic and Lower Jurassic strata in the district, comparing the previous nomenclature with that adopted for the Birmingham district.

System/Series		Stage	Lithostratigraphy		
			Previous nomenclature (Hull, 1869; Eastwood et al., 1925)	Nomenclature used in this account (based on Warrington et al., 1980)	
				Group	*Formation*
Lower Jurassic		Hettangian	Lower Lias	Lias Group[†]	Blue Lias Formation[†]
Triassic	Upper	Rhaetian	Rhaetic	Penarth Group	Lilstock Formation (Cotham Member*)
					Westbury Formation
			Tea Green Marl	Mercia Mudstone Group	Blue Anchor Formation
		Norian	Keuper Marl		
		Carnian			Arden Sandstone Formation
	Middle	Ladinian			
		Anisian	Lower Keuper Sandstone		Bromsgrove Sandstone Formation
	Scythian	Induan – Olenekian	Upper Mottled Sandstone	Sherwood Sandstone Group	Wildmoor Sandstone Formation
			Bunter Pebble Beds		Kidderminster Formation
			Hopwas/Quartzite/Sutton breccias & Barr Beacon Beds		Hopwas Breccia

* only the Cotham Member is represented in this district

† the basal 4 to 5 m of the Lias Group is of Rhaetian age (p.80).

Triassic succession, was formerly regarded as Permian to Triassic in age, but the results of recent palaeomagnetic (Johnson, 1995) and provenance studies (Hallsworth, 1994) indicate that it is genetically related to the overlying known Triassic rocks. It is overlain, in upward sequence, by the Sherwood Sandstone, Mercia Mudstone and Penarth groups (Warrington et al., 1980). The last represents the deposits of brackish and marine depositional environments introduced by the major Late Triassic marine transgression (Warrington and Ivimey-Cook, 1992).

HOPWAS BRECCIA

The Hopwas Breccia is up to 10 m thick and consists of yellow-brown to red-brown, medium- to coarse-grained, pebbly sandstone and breccia. Pebble clasts are generally subangular to subrounded and include a wide variety of lithologies derived from local and more distant sources. They include quartzite, red, grey and yellow calcareous and decalcified sandstone (including highly fossiliferous Silurian sandstone), pale grey Carboniferous limestone and rotted igneous rocks. The proportion and composition of the clasts varies from one locality to another, so that in the vicinity of Barr Beacon [05 96], in the north of the district, quartzite and Silurian sandstone clasts are common, with a paucity of limestone clasts, whereas a few kilometres to the north of the district, at the type Hopwas area [SK 17 05], Carboniferous limestone commonly forms up to 40 per cent of the clasts (Eastwood et al., 1925). This is in accord with the overall southerly decrease in the proportion of Carboniferous limestone pebbles in the unit within the district (Edwards et al., 1925). In the south of the district, near Tessall Lane [006 785] the unit is locally known as the Quartzite Breccia (Wills and Shotton, 1938; Old et al., 1991) and is composed almost entirely of angular clasts locally derived from the Lickey Quartzite, with only a sparse sandstone or claystone matrix.

The Hopwas Breccia cannot be traced throughout the district. It is overstepped locally by the Kidderminster Formation and although exposures are scarce, this lateral impersistence probably reflects deposition and preservation of the Hopwas Breccia within topographical lows on the early Triassic landscape. The outcrop pattern, the presence of various locally derived pebble clasts at different localities, and bedforms (see below) indicate deposition as both proximal and distal alluvial fans, sourced from an emergent Palaeozoic hinterland.

The age of the Hopwas Breccia is poorly defined; no fossils are known from the formation. Smith et al. (1974, table V) considered the unit and its correlatives to be early Permian in age, possibly equivalent to the aeolian Brignorth Sandstone present to the north- and south-west, in the Stafford and Worcester basins (Barclay et al., 1997). The Hopwas Breccia overlies Upper Carboniferous rocks (Salop Formation) with a slight angular unconformity in the district, but is overlain disconformably by the basal conglomerate of the Kidderminster Formation, regarded as Early Triassic in age (Warrington

et al., 1980). Palaeomagnetic studies at Barr Beacon and localities to the north of the district, at Hopwas, and in boreholes at Shenstone (Johnson, 1995) indicate that the Hopwas Breccia has a palaeomagnetic signature consistent with a Triassic palaeolatitude position of about 20° north of the equator, which suggests a Triassic age is more likely.

The Quartzite Breccia is regarded here as broadly coeval with the Hopwas Breccia. However, Old et al. (1991) included it within the Sherwood Sandstone Group, although the base of that group has been taken traditionally at the base of the overlying Kidderminster Formation (Warrington et al., 1980). Here, the Hopwas Breccia is not included in the Sherwood Sandstone Group, since it was deposited in a markedly different environment, reflected in its distinctive bedforms and provenance. Deposition of the Hopwas Breccia was followed by basinwide subsidence and the deposition of conglomerate containing far-travelled exotic pebbles (the Kidderminster Formation at the base of the Sherwood Sandstone Group).

The best exposures are in the disused quarries below the north-trending Barr Beacon Ridge [0603 9688] where the unit was formerly named the Barr Beacon Beds (Landon, 1890). It unconformably overlies the Upper Carboniferous Salop Formation (Enville Member), and is disconformably succeeded by the scarp-forming Kidderminster Formation. At Pinfold Lane Quarry [0592 9672 to 0690 9662] about 7 m of the Hopwas Breccia is exposed (Plate 5). It comprises orange-brown, medium- to coarse-grained sandstone in 0.2 to 1 m thick, laterally wedging, tabular sets, internally structured by planar and low-angle, trough cross-bedding. Thin, red-brown mudstone partings and mudstone rip-up clasts are present. Subangular pebbles, 20 to 50 mm in diameter, occur mostly as lag deposits lining erosively based sets and less commonly along the foresets. They define upward-fining units, so that a basal pebble lag passes upward through trough or planar cross-bedding with sparse pebbles or granules lining the foresets, to planar or wavy bedded, parallel-laminated mudstone partings containing sparse desiccation cracks. A fluvial-dominated, distal alluvial fan setting is envisaged for this locality. Sandstone and quartzite clasts are most common, with subordinate, highly weathered, volcanic rocks. A yellow decalcified sandstone pebble yielded a prolific shelly fauna including brachiopods (*Stricklandia* cf. *lens*) and solitary corals (*Rhabdocyclus* sp.), considered to be Llandovery (early Silurian) in age (Rushton, 1994). This suggests provenance from either the Welsh Borderlands, or from the Rubery Sandstone which crops out in a fault-bounded wedge along the Great Barr Fault, about 2 km to the south-west. White crinoidal chert pebbles, probably derived from the Carboniferous limestone or reworked from the calcareous conglomerate beds of the Enville Member, have been recorded (Eastwood et al., 1925). The boundary with the overlying Kidderminster Formation is sharp; erosive scours at the base of the conglomerate are infilled locally with typical well-rounded 'Bunter' pebbles.

East of Sutton Coldfield, the Hopwas Breccia forms a poorly defined, narrow outcrop at Sutton Park [109 966],

Plate 5 Basal conglomerate of the Triassic Kidderminster Formation (base marked B) overlying the Hopwas Breccia. Pinfold Lane Quarry [069 966], near Barr Beacon (GS 709).

though it is mostly overstepped by the Kidderminster Formation in this area. At Keeper's Pool Quarry [1075 9663], former exposures showed that the upper boundary of the Hopwas Breccia is sharp but highly irregular (Barrow et al., 1919). Subangular clasts consist of grey quartzite (43 per cent), decomposed feldspathic igneous rocks (37 per cent) and sandstone (20 per cent) (Boulton, 1933). Nearby, at Blackroot Pool [1091 9705] the following pebble count was given by Eastwood et al. (1925) — decomposed feldspar-bearing igneous rocks (35 per cent), Silurian 'grits' and sandstone (30 per cent), quartzite (30 per cent) and Carboniferous chert (5 per cent); some of the pebbles showed slight facetting, attributed to wind abrasion. Boulton (1924, 1933) concluded that these 'Sutton breccias' showed close affinity with the Nechells Breccia, proved in boreholes in central Birmingham, and furthermore considered these beds as correlatives of the Permian Clent (Breccia) Formation. However, petrographical, heavy mineral, and palaeomagnetic studies (Hallsworth, 1994; Johnson, 1995), and field relationships, support the retention of the Hopwas Breccia (including the locally named Barr Beacon Beds, Sutton Breccia and Quartzite Breccia) as a separate unit of probable Triassic age, stratigraphically younger than the Permian Clent Formation and separated from the overlying Kidderminster Formation by a disconformity.

South-west of Great Barr Hall, the Hopwas Breccia wedges out so that the Kidderminster Formation oversteps unconformably onto the Enville Member. It crops out again on a ridge from Perry Hill to Tower Hill, near Hamstead, where former exposures in road and canal cuttings revealed up to 9 m of sandstone with

'marly breccia' containing clasts of quartzite (possibly Lickey Quartzite), quartz and sandstone (Eastwood et al., 1925). Near Park Farm [030 915] the Hopwas Breccia was formerly exposed in small excavations, now obscured, and reflected by angular pebbles in the soil (Eastwood et al., 1925, p.71). South of here the outcrop is conjectural due to the cover of drift deposits. The unit is absent below the basal Kidderminster Formation in Warley Park [008 850], and has only been observed in the south-west of the district in the Tessall Lane area [006 785], described above, where the breccia is entirely composed of locally derived Cambrian Lickey Quartzite (Old et al., 1991).

In central Birmingham, breccia beds have been proved below the Kidderminster Formation in a number of boreholes). There has been much debate about the stratigraphy of these beds (Boulton, 1924, 1933, 1970a, 1976) which have variously been included in the Hopwas Breccia (or equivalents), Clent Formation (including the coeval Nechells, Warley and Northfields breccias) or the Enville Member. Difficulty in distinguishing the Hopwas Breccia is no doubt due to the locally variable clast composition, and the probable presence of reworked clasts, such as Carboniferous limestone, derived from conglomerates of the Enville Member. However, in a number of well recorded boreholes (for example Nechells Gasworks, Windsor Street Gasworks No. 3, H P Sauce; p.113), between 4 and 37 m of Hopwas Breccia were proved below the Kidderminster Formation, and resting on sandy breccias referred to the Clent Formation (Eastwood et al., 1925; Boulton, 1933; Butler and Lee, 1943). The breccia is characterised by large angular quartzite, and siliceous sandstone pebbles (up to 0.35 m long axis), and interbeds of red mudstone ('marl') and red sandstone. The high proportion (up to 90 per cent) of quartzite clasts, probably derived from the Lickey Quartzite, in these boreholes indicates a provenance affinity with the Tessall Lane outcrop in the south of the district (Wills and Shotton, 1936). In contrast to the underlying Clent Formation, igneous clasts are absent, and the quartzite clasts also lack the characteristic haematitic patina of the Clent Formation clasts.

The heavy mineral assemblages of the Hopwas Breccia at Barr Beacon were analysed in a regional study of the mineralogy of the Carboniferous, Permian and Triassic rocks of the area (Hallsworth, 1994). Pioneering work on the heavy mineral assemblages (Fleet, 1923, 1925,1927) was hampered by the, then, poor understanding of the affects of variations in hydraulic conditions at the time of deposition, and of dissolution during diagenesis, leading to incorrect interpretation. These factors can now be mitigated by determining provenance-sensitive heavy mineral ratios, such as monazite: zircon (MZi), and staurolite: rutile (SRi) (Hallsworth, 1994).

Samples from Pinfold Lane Quarry, Barr Beacon, yielded a heavy mineral suite dominated by zircon, with moderate amounts of tourmaline and rutile; garnet, anatase, monazite and staurolite are minor constituents. The absence of apatite, commonly found in these beds at other localities in the area (Hallsworth, 1994), is probably due to weathering at outcrop. The MZi values (3.4–7.8) and SRi values (16.2–36.1) are similar to those from the type Hopwas area, suggesting a similar provenance. These values are very much lower than those obtained from samples of the Kidderminster Formation at the same locality, suggesting that the formations were sourced from different provenances. This supports the exclusion of the Hopwas Breccia from the Sherwood Sandstone Group. Moderate amounts of monazite and staurolite in the samples indicates derivation from amphibolite metasediments, either in the first erosional cycle or through reworking. Very high values for these mineral ratios in the overlying Sherwood Sandstone have been attributed to a far-distant southerly source, probably the Armorican Massif (Hallsworth, 1994). If this area also contributed to the Hopwas sediments, then it was either heavily diluted by a more local source of sediment (Lower and Upper Palaeozoic rocks) poor in these minerals, or much of the sediment may have been locally reworked from the underlying Enville Member, which has a lower proportion of these minerals. The latter hypothesis seems likely since the Hopwas Breccia locally contains jasper, chert and Carboniferous limestone clasts which may be secondarily reworked from the calcareous conglomerates of the Enville Member.

SHERWOOD SANDSTONE GROUP

The Sherwood Sandstone Group comprises continental, arenaceous strata of Early and Mid-Triassic age in Britain (Warrington et al., 1980). In the central Midlands region the group is subdivided, in upward sequence, into the Kidderminster Formation, Wildmoor Sandstone Forma-

tion and Bromsgrove Sandstone Formation (Warrington et al., 1980); these units (Table 14) are broadly equivalent to the 'Bunter Pebble Beds', 'Upper Mottled Sandstone' and 'Lower Keuper Sandstone', respectively of earlier nomenclature (Hull, 1869; Eastwood et al., 1925).

Regionally, the group was deposited in a series of rapidly subsiding, generally north–south-orientated basins, such as the Knowle Basin in this district (Wills, 1956), separated by horsts or blocks, represented, in this district, by the South Staffordshire and the Warwickshire coalfields. The tectonic regime was one of east–west extension with listric growth faulting along major bounding faults such as the Western Boundary Fault of the Warwickshire Coalfield and the Birmingham Fault (Chadwick and Smith, 1988). Triassic sedimentation extended onto the highs defined by these major bounding faults, but the succession is generally thinner in these areas.

Rapid subsidence of the Knowle Basin coincided with uplift of the Armorican hinterland far to the south. This resulted in high sediment flux (discharge) during which pebble beds and sand were deposited in a predominantly fluvial environment (Wills, 1956; Audley-Charles, 1970).

In the Knowle Basin, the base of the Sherwood Sandstone Group has not been proved in boreholes, but the Blyth Bridge Borehole [2119 8979] proved 760 m of sandstone and pebbly sandstone beds assigned to the Sherwood Sandstone Group. Geophysical evidence suggests a minimum thickness of about 625 m (Allsop, 1981), and there is geophysical evidence of growth faulting and rapid thickening of some stratigraphical units across faults, such as the Meriden Fault which was active during deposition (Chapter Nine; Allsop, 1981; Old et al., 1990). The age of the Sherwood Sandstone Group is imprecisely known in the district, but comparison with adjacent areas suggests it ranges from Early Triassic (Induan–Olenekian) to early Mid-Triassic (Anisian) (Benton et al., 1994; Barclay et al., 1997).

Kidderminster Formation

The Kidderminster Formation (Warrington et al., 1980) crops out intermittently along a broad, predominantly drift-covered tract to the north-west of the Birmingham Fault (Figure 1). It consists of pebble conglomerate, pebbly sandstone and medium- to coarse-grained, cross-bedded sandstone with sparse, thin mudstone beds, and ranges in thickness from 45 to 120 m. The Kidderminster Formation rests disconformably on the Hopwas Breccia, or, where the breccia is absent, unconformably on Upper Carboniferous and Lower Permian rocks (Salop Formation). Just south of the district, in the Lickey Hills, it rests unconformably on Lower Palaeozoic rocks (Old et al., 1991). The boundary with the overlying Wildmoor Sandstone Formation is gradational, except in the Frankley area, in the south-west of the district, where that unit is absent and the Kidderminster Formation is overstepped by the Bromsgrove Sandstone.

Throughout most of the outcrop, and in boreholes in central Birmingham, the lower part of the formation consists of well-rounded, clast-supported, pebble conglomerate with a matrix and interbeds of subordinate medium-grained, micaceous sandstone ('Bunter Pebble Beds'; Table 14). The formation is generally very friable and weakly cemented, but locally, harder calcite-cemented beds are present. Petrographical studies of the clasts (Lamont, 1940, 1946, 1948; Campbell Smith, 1963, Wills, 1956), the heavy minerals (Fleet, 1923, 1925, 1927; Hallsworth, 1994) and an isotopic study of detrital mica in adjacent areas (Fitch et al., 1966), indicate a southerly provenance. Pink-grey ('liver-coloured') quartzite, and vein-quartz make up about 80 to 90 per cent of the pebbles; the remainder comprise a varied assemblage of igneous, metamorphic and sedimentary rocks. Quartzite pebbles have characteristic pressure solution pits at points of contact in these clast-supported beds. Sparse, well-rounded ('millet seed') grains of probable aeolian origin, described from adjacent areas (Old et al., 1991), may have been reworked from Permian rocks outside the district, or were blown into the area from coeval aeolian dunes.

Clast composition gives some indication of the source area. Silicified limestone pebbles containing rugose corals and brachiopods, collected from the Kidderminster Formation in Sutton Park, were derived from the Carboniferous limestone (Eastwood et al., 1925, p.75). Lamont (1946) recorded fossiliferous Ordovician, Silurian and Devonian pebbles from the district, and concluded that the Silurian (upper Llandovery) pebbles were derived from the West Midlands or the Welsh Borderlands; the source area for the Ordovician and Devonian rocks was probably the Armorican highlands, represented by present-day Brittany (Grès de Mai and Grès Armoricain) or Cornwall (Goran Haven). Campbell Smith (1963) also concluded that the distinctive igneous rocks, including foliated granite and quartz-porphyry rich in tourmaline, were derived from a similar southern source area. More local sources, such as a Variscan klippe or Permian breccias, now buried beneath Mesozoic cover rocks, were suggested by Wills (1956). However, a more distant provenance is supported by the isotopic dating of detrital micas which indicate derivation, in part, from crystalline igneous rocks with ages in the range 200 to 300 Ma (Fitch et al., 1966). This is further supported by local (Hallsworth, 1994) and regional (Morton, 1992) studies of heavy mineral suites.

In addition, regional palaeocurrent and isopachyte data (Fitch et al., 1966; Audley-Charles, 1970) indicate deposition of the Kidderminster Formation from a major, northward-flowing, braided river system ('Budleighensis River' of Wills, 1956), sourced predominantly from the Variscan highlands located far to the south of the district (Warrington and Ivimey-Cook, 1992), but with minor, local input from rivers draining the Variscan highs of the west Midlands. At the time of deposition of the basal conglomerate, the Late Permian topography had been partially subdued by local alluvial fan sedimentation (Hopwas Breccia). The basin-wide event represented by high sediment flux (discharge) manifest in the basal pebble conglomerate of the Kidderminster Formation was probably due to rapid tectonic uplift of the source area, and subsidence of the Worcester and Knowle basins along extensional, north-trending, synsedimentary faults (Holloway, 1985; Chadwick and Smith, 1988). The pebble conglomerate facies gradually onlapped onto intervening highs, such as the eastern margin of the South Staffordshire Coalfield in the west of the district.

Two or three cycles of upward-fining conglomerate-sandstone (about 10 to 20 m thick) are present in the lower part of the Kidderminster Formation, but the upper part is dominated, in this district, by upward-fining sandstone cycles. These consist of red-brown to yellow-brown, fine- to coarse-grained, micaceous, feldspathic, trough cross-bedded sandstone with scattered small pebbles and pebble lenses (channel lag deposits); clasts are similar to the suite found in the basal conglomerate beds. Some of the cycles are capped by thin beds of parallel-bedded or blocky red mudstone; the latter is also present as angular mudstone rip-clasts in some of the channel sandstones, indicating penecontemporaneous erosion of the floodplain. Wills (1970a, 1976) recognised three 'miocyclothems' (I–III) and part of a fourth (IV) in this formation in the Birmingham district.

A fluvial depositional environment is indicated for the Kidderminster Formation by the well-rounded, far-travelled pebble clasts, characteristic bedforms, upward-fining rhythms, sandstone petrology and paucity of fauna. The pebble conglomerate beds probably represent longitudinal or transverse bars deposited in a braided river channel. Large-scale, trough cross-bedded sandstones with pebble lags, in the upper part of the unit, suggest laterally migrating channels, with mudstone beds deposited from suspension under conditions of waning flow. An overall upward decrease in maximum pebble size, and a relative increase in the proportion of sandstone to conglomerate through the formation, suggest a decrease in stream velocity through time. This may be equivalent to the first-order geomorphological cycles of Schumm (1977, 1993). However, the presence of higher-order, upward-fining conglomerate–sandstone cycles (10–20 m thick) in the lower part of the formation indicates marked variations in stream discharge and stream velocity, in an environment characterised by ephemeral streams. These streams, possibly resulted from flood events initiated in the mountainous hinterland, in a semi-arid climate (Audley-Charles, 1970; Warrington and Ivimey-Cook, 1992).

No fossils have been found in the Kidderminster Formation in the district. However, records of crustaceans (*Euestheria*) near Walsall, in the Lichfield district (Barrow et al., 1919), and *Permichnium völckeri*, a trace fossil attributed to insects, in the Droitwich district (Wills and Sarjeant, 1970) testify to the existence of a fauna in the region during Early Triassic times. Wills and Sarjeant (1970) and Sarjeant (1996) inferred the presence of several groups of reptiles, on the basis of footprints from the Droitwich district, but King and Benton (1996) have reinterpreted these structures as inorganic in origin.

The Kidderminster Formation crops out over a wide area west of Sutton Coldfield. Exposures in the lower conglomerate-rich part of the formation, at Blackroot Pool Quarry [1091 9705] and Keeper's Pool Quarry

[1075 9663] in Sutton Park, reveal thin beds of weakly cemented pebbly sandstone and lenticular beds of medium-grained sandstone, containing well-rounded pebbles and, locally, cobbles up to 0.46 m in diameter. Clasts consist mostly of quartzite but include igneous rocks, Silurian sandstone, and white fossiliferous chert derived from the Carboniferous limestone (see Eastwood et al., 1925, p.75 for a list of fossils).

Farther west, the lower part of the Kidderminster Formation overlies the Hopwas Breccia with a planar erosional boundary. It is well exposed in quarries along the Barr Beacon escarpment [0592 9672 and 9690 9662] and consists of pebble- to cobble-grade, clast-supported conglomerate, with a matrix and thin lenses of red to yellow-brown, coarse- to medium-grained sandstone. The clasts consist of 'liver-coloured' quartzite (up to 0.25 m diameter), vein quartz and subordinate quartz felsite, weathered granitic rocks, and Silurian sandstone and limestone. The conglomerate passes up into pebbly sandstone-dominated beds which form the low ground to the east of Barr Beacon; up to 9.5 m of red and white mottled, medium-grained, trough and planar cross-bedded sandstone, locally with mudstone rip-up clasts, was exposed in Bliss Sand and Gravel Pit [0692 9999].

In the Perry Barr and Queslett areas, the Kidderminster Formation oversteps the Hopwas Breccia to overlie the Salop Formation (Hardaker, 1912). Small exposures in the back-face of former quarries [063 944] showed up to 4 m of typical conglomerate lithologies passing up to cross-bedded sandstone with pebble lenses.

The formation gives rise to a prominent scarp feature from Hamstead Hill [050 917] to Handsworth Hall [045 908], where a road cutting [0487 9188] exposed red, pebbly, micaceous sandstone with lenses of well-rounded, clast-supported pebble conglomerate. South of here, in the Handsworth area, the Kidderminster Formation is mostly obscured by drift. However, the lower part was exposed in the Birmingham Metro cutting where up to 4 m of red-orange and yellow, medium- to coarse-grained, trough cross-bedded sandstone with quartz and quartzite pebble lag lenses unconformably overlies the Salop Formation.

The lower part of the formation is exposed in small pits and quarries in the drift-free areas which form the high ground around Warley Park. Disused gravel pits [008 8598 and 0088 8614] along the scarp ridge expose pebbly soil with the typical suite of well-rounded clasts. Up to 4.5 m of red-brown, pebbly sandstone and pebble conglomerate were reported from a disused quarry (Wills, in Eastwood et al., 1925), and similar lithologies were formerly exposed in gravel pits [0092 8722, 0090 8737, 0165 8865] to the north of Thimblemill Brook.

There are few exposures in the Harborne area, due to the cover of drift, but red, fine-grained sandstone with common rounded quartz grains was seen below the alluvium in Bourn Brook [019 832]. Farther south, in the Longbridge area, the basal conglomerate lithofacies found in adjacent areas is absent and the formation, exposed in Merritt's Brook [0091 7992, 00957973], consists of red, fine- to coarse-grained, soft, micaceous sandstone with small quartz pebbles and subordinate beds of red mudstone; trough and planar cross-bedding are typical. In this area, the Wildmoor Sandstone is absent and the Kidderminster Formation is overstepped by the Bromsgrove Sandstone. In Tessall Lane [0098 7848], Wills and Shotton (1938, p.181, pl. 1) recorded soft 'Bunter' sandstone overlying the Hopwas (Quartzite) Breccia, and a disused gravel pit [0079 7879] formerly exposed 1.5 m of red and red-brown, micaceous sandstone with small quartz pebbles.

In the central Birmingham area, the Kidderminster Formation has been proved up to 119 m thick in many water boreholes (Eastwood et al., 1925; Butler and Lee, 1943; Wills, 1976). However, the lack of lithological contrast within the Sherwood Sandstone Group makes it difficult to determine, with any certainty, the boundaries, and hence the true thickness of the formations in the majority of published borehole records. This is particularly so with the upper boundary of the Kidderminster Formation, since the overlying Wildmoor Sandstone is lithologically similar to the upper, sandstone-dominated part of the Kidderminster Formation. Furthermore, in the area immediately north-west of the Birmingham Fault the lithology of the lower part of the Kidderminster Formation is atypical in that it has a small proportion of rounded pebbles, and even subangular clasts have been reported (Boulton, 1933; Butler and Lee, 1943). In boreholes located about 1 km north-west of the Birmingham Fault the Kidderminster Formation is about 76 m thick (Butler and Lee, 1943); it thins south-eastwards towards the fault and is about 12 m thick in the H P Sauce (Midland Vinegar) Borehole and the Windsor Street Gasworks Borehole. Wills (1976) considered that the formation is absent in the Mitchell and Butler City Road Borehole and he proposed the term 'City Road Beds' for the non-conglomeratic sequence overlying the basal 'pebble beds'; this unit is probably equivalent to both the upper part of the Kidderminster Formation and the lower part of the overlying Wildmoor Sandstone. However, this apparent thinning of the Kidderminster Formation is probably the result of a lateral change from pebble-rich to pebble-poor sandstone towards the Birmingham Fault, making it difficult to distinguish the Kidderminster Formation from the overlying Wildmoor Sandstone. To the south-east of the Birmingham Fault, the basal conglomeratic lithofacies is again present; in this area the formation ranges in thickness from 68 m in the Nechells Gasworks Borehole No. 3 to 115 m in the Smart and Sons Borehole. Over much of the Birmingham City area located to the south-east of the Birmingham Fault, which represents the western margin of the Knowle Basin, the Kidderminster Formation was thought to be unconformably overlain by the Bromsgrove Sandstone (Wills, 1976), but in a few boreholes the Wildmoor Sandstone may also be present (see below). The Blyth Bridge Borehole [2119 8979], located in the east of the Knowle Basin, proved a 578 m-thick succession of sandstone and pebbly sandstone below the Bromsgrove Sandstone (Figure 21). It is uncertain whether this sequence represents only the Kidderminster Formation or includes the overlying Wildmoor Sandstone Formation, and consequently the formations are shown as 'undifferentiated' on the cross-section on the 1:50 000 Series map (Birmingham Sheet 168).

The Kidderminster Formation, sampled at Barr Beacon and from the Hopwas-Hints area, located to the north in the Lichfield district (Sheet 154), yielded a heavy mineral suite rich in staurolite, zircon and tourmaline, with moderate amounts of rutile and monazite (Hallsworth, 1994). The most distinctive heavy mineral ratios (see section under Hopwas Breccia for explanation) are very high SRi values (61.1–84.1) and moderately high MZi values (7.4–17.7); these high ratios are typical of regional values (Morton, 1992) and serve to distinguish the formation from the Hopwas Breccia. There is an overall decline in MZi ratios from south to north in Britain, with the district representing an intermediate station (Morton, 1992). This is consistent with a northward dilution of monazite from a source in the Armorican area. This trend suggests that in areas north of the Stafford–Needwood–Knowle basins, of which the district forms a part, the sediments were derived from areas poor in monazite, such as the Palaeozoic hinterland areas on the east and west sides of the basin (Chapter Nine). The distribution patterns for staurolite are less certain due to dissolution of this mineral during burial. However, the broadly coeval Budleigh Salterton Pebble Beds on the margin of the Wessex Basin, to the south, are also rich in staurolite and monazite, suggesting a similar source area, believed to be the amphibolite facies metasediments found in the Armorican Massif (Hallsworth, 1994).

Wildmoor Sandstone Formation

The Wildmoor Sandstone Formation (Warrington et al., 1980) is equivalent to the former 'Upper Mottled Sandstone'. It consists predominantly of orange-red, fine-grained, micaceous, soft sandstone and subordinate, thin beds of red-brown and green-grey mudstone. It crops out to the north-west of the Birmingham Fault, but is generally overlain by drift deposits, and in the north-east and south-west parts of the outcrop it is overstepped by the Bromsgrove Sandstone Formation. The maximum thickness is about 120 m proved in boreholes to the north-east of the Birmingham Fault, but locally in north and south Birmingham it is absent. The formation was considered to be absent to the east of the Birmingham Fault (Wills, 1976), but the 578 m-thick sandstone succession proved beneath the Bromsgrove Sandstone in the Blyth Bridge Borehole, located in the east of the Knowle Basin, may be attributed, in part, to the Wildmoor Formation. The fine grain size and soft, poorly cemented nature of the sandstone favoured its exploitation as a moulding sand for use in the foundries in central Birmingham, where it was known locally as the 'Hockley moulding sand' (Boswell, 1919). The formation is equivalent to the upper part of 'miocyclothem BSIV' of Wills (1976).

The lower boundary of the Wildmoor Sandstone is gradational; it rests conformably on the Kidderminster Formation, and is distinguished from the latter by its fine grain size, typical 'foxy red' colour, and mottled appearance due to the presence of weathered pale grey feldspar grains. The upper boundary is more easily distinguished at the base of the unconformable Bromsgrove Sandstone, which is characterised by red-brown, quartzose, calcareous- and ferruginous-cemented, pebbly sandstone.

Moulding sand worked in the pit at Key Hill Cemetery, Hockley, gave the following analysis (Boswell, 1919): silica 85 per cent, alumina 6.75 per cent and iron oxides 1.5 per cent; the remainder comprises small quantities of various substances of which potassium carbonate is abundant. Grain-size analyses shows 40 to 67 per cent medium sand and 28 to 46 per cent fine sand, with less than 2 per cent coarse sand. The characteristic 'foxy red' colour is due to a thin pellicle of iron oxide on the quartz grains.

Bedforms typically comprise tabular or low-angle, laterally wedging cosets, about 0.3 to 0.5 m thick, internally structured by low-angle, planar cross-bedding or gently asymptotic trough cross-bedding; small granules and red mudstone rip-up clasts are present in places above erosively scoured surfaces. Palaeocurrent directions derived from both trough and planar cross-bedding indicate a unimodal palaeoflow towards the north-west (Hassan, 1964).

The heavy mineral suite is similar to the Kidderminster Formation and is dominated by abundant tourmaline and zircon, with apatite and anatase, and subordinate amounts of feldspar, garnet, rutile, barytes, staurolite, ilmenite and magnetite (Fleet, 1923). The low proportion of staurolite, in comparison with the underlying formation, may indicate that reworking of the Kidderminster Formation occurred, since staurolite is relatively unstable and would be the first mineral to disappear during recycling.

Fossils have not been found in the Wildmoor Sandstone in the district. Trace fossils have, however, been reported from boreholes farther south-west and west, at Bellington (Wills, 1970b; Wills and Sarjeant, 1970; Pollard, 1985), and Churchill (Wills, 1976, pp.55, 56), in the Droitwich and Dudley districts, respectively. These comprise invertebrate traces (*Isopodichnus* and *Planolites*) and vertebrate footprints, associated with rain pits and ripple marks; the footprints have been reinterpreted as structures of inorganic origin (King and Benton, 1996). A fragmentary fish, possibly a member of a group (the perleidids) that had marine affinity, was recorded near Kidderminster (White, 1950).

The petrographical characteristics, such as the fine grain size and paucity of pebbles, together with large-scale, low-angle cross-bedding and a low proportion of mudstone beds (interpreted as floodplain overbank deposits), suggest deposition in a fluvial environment. This may have been a distal braidplain, characterised by laterally migrating rivers in which flow rates were lower than during the deposition of the underlying Kidderminster Formation. Sediment flux remained high, but the general absence of exotic pebble clasts suggests that sediment was reworked from the upper part of the Kidderminster Formation; this is supported by the heavy mineral data (Fleet, 1923). Upward-fining cycles with mudstone tops, and the presence of vertebrate trace fossils, suggest deposition from ephemeral streams. The presence of rounded grains may be due to primary aeolian deposition from surrounding dune fields, or reworking of aeolian deposits, such as the Permian Bridgnorth Sandstone which crops out to the west of the district.

There are few natural exposures in the Wildmoor Sandstone in the district, due to the extensive drift cover and urbanisation; most exposures are found in partly degraded moulding sand pits in central Birmingham and recent excavations associated with the Birmingham Metro. Details of exposures that existed at the time of the previous survey are given in Eastwood et al. (1925).

Typical red, micaceous, fine-grained sandstone was exposed in the Botanical Gardens, Chad Valley [0486 8532], and about 5 m of red, fine-grained sandstone with marl beds were formerly seen, below the drift, in quarries for moulding sand [0384 926, 021 874] near Handsworth.

In central Birmingham, the 'Hockley moulding sand' was formerly worked for foundry sand. The largest quarry [0598 8815 to 0600 8806], now Key Hill Cemetery, was adjacent to the Birmingham Mint. The quarry face is now partially obscured, but up to 15 m of orange-brown, fine-grained, feldspathic, slightly micaceous sandstone was exposed below about 5 m of glaciofluvial deposits; large-scale, low-angle, planar and slightly asymptotic trough cross-bedding is common. Nearby, temporary excavations for the Birmingham Metro revealed similar trough cross-bedded sandstones in tabular sets (0.2 to 0.3 m thick); palaeocurrent flow towards the east was determined from the cross-bedding. The formation was exposed in trial pits located along the Metro track, from Northwood Street to Norton Street [0521 8849]; at the latter locality up to 4 m of red, and orange-brown, fine-grained, micaceous, thin-bedded, feldspathic sandstone with thin beds of red mudstone was exposed below the pebbly drift. Moulding sand was formerly exploited to a depth of 36 m, below a cover of glaciofluvial deposits at the Park Road Sand Pit [0531 8854] and in small pits at Summer Hill Road [0565 8730, 0583 8717]; it was formerly seen in excavations for Snow Hill Station (Eastwood et al., 1925), close to the boundary with the overlying Bromsgrove Sandstone. The upper part of the formation also crops out near Aston Park where it gives rise to sandy soil and was formerly seen in small exposures to the west of the park [078 898, 0745 8993].

No outcrops were seen in the Harborne area during the present survey; Eastwood et al. (1925) recorded 15 m of typical lithologies in quarries adjacent to the railway at Selly Oak [043 830]. South of here, the formation is over-stepped by the Bromsgrove Sandstone. Boreholes in the Harborne area proved up to 100 m of red-orange, soft sandstone with thin beds of red mudstone and scattered granule or pebble stringers.

In central Birmingham, to the north-west of the Birmingham Fault, the Wildmoor Sandstone ranges in thickness from 30 to 86 m in boreholes (Eastwood et al., 1925; Butler and Lee, 1942). To the south-east of the fault, it was formerly regarded as absent (Eastwood et al., 1925, p.80; Butler and Lee, 1942), although red sandstone and thin mudstone beds typical of the formation have been recorded in the Stechford area (Piper, 1993b) where the Kitt's Green Borehole [1481 8755], Southalls No. 4 Borehole (Wills, 1976) and Bromford Tube Co. (b) Borehole [1131 9895] proved about 34 m, 18 m and 7 m, respectively. In the Erdington area up to 39.4 m of beds assigned to the Wildmoor Sandstone were proved in the Vickers-Armstrong Borehole [1289 9062]. Thus, boreholes located a few kilometres to the south-east of the Birmingham Fault indicate that the formation may be locally preserved in parts of the Knowle Basin, below the unconformable Bromsgrove Sandstone (Wills, 1976, fig. 4). However, the boundary with the underlying Kidderminster Formation is not distinct. For example, the Blyth Bridge Borehole [2119 8979], located in the east of the Knowle Basin, proved about 578 m of sandstone and pebbly sandstone, below the Bromsgrove Sandstone; the upper part of this sequence may be equivalent to the Wildmoor Formation.

Bromsgrove Sandstone Formation

This formation, formerly known as the 'Lower Keuper Sandstone', forms a narrow outcrop to the north-west of the Birmingham Fault, and has been proved, below the Mercia Mudstone, in many boreholes to the south-east of the fault; small fault-bounded outcrops also occur in the south-east near Meriden. The Bromsgrove Sandstone (Warrington et al., 1980) ranges in thickness from about 84 to 180 m in the district. Isopach maps (Figure 20 and Wills, 1976, fig. 11) show it to be thickest in the Saltley area on the downthrow side of the Birmingham Fault, and to thin from there to the south and south-west; about 84 m were proved in the Longbridge Laundry Borehole. In the east of the Knowle Basin 182 m were proved in the Blythe Bridge Borehole, but only 79 m the nearby Dumble Farm Borehole (Figure 21); the greater thickness in the former is attributed to growth faulting on the Maxstoke Fault, although part of the formation may be faulted out in the Dumble Farm Borehole (Old, 1989).

In adjacent areas and in some boreholes in the district (Wills, 1976), the formation has been subdivided into three members; these comprise, in upward sequence, the 'Basement Beds', 'Building Stones' and 'Waterstones' (Hull, 1869) which have been formalised as the Burcot, Finstall and Sugarbrook members, respectively (Old et al., 1991). These subdivisions have not been mapped in this district, and are not readily distinguished in boreholes (Figure 22). The base of the Bromsgrove Sandstone is generally sharp and is marked by a strong topographical scarp feature; the formation rests unconformably on, and oversteps, the Wildmoor Sandstone, Kidderminster Formation and Clent Breccia. In boreholes located a few kilometres to the south-east of the Birmingham Fault, the Bromsgrove Sandstone generally rests unconformably on the Kidderminster Formation, but locally a thin representative of the Wildmoor Sandstone is present (see above). The unconformity is correlated with the Hardegsen Unconformity (Geiger and Hopping, 1968) and represents a period of

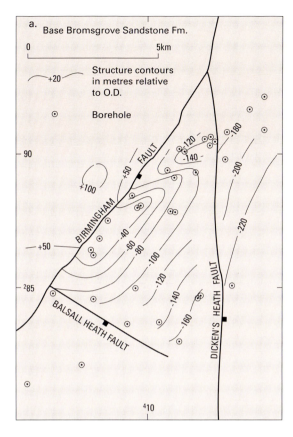

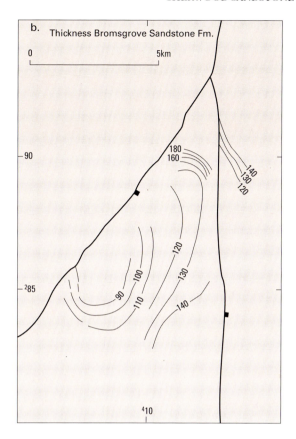

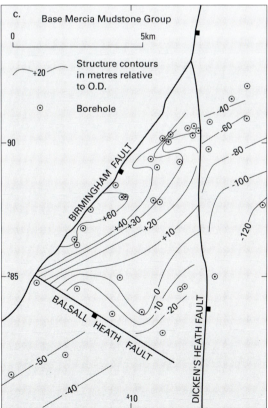

Figure 20 Structural and thickness maps for the Triassic strata in central Birmingham.

a. Contour map of the base of the Bromsgrove Sandstone Formation.
b. Isopachyte map for the Bromsgrove Sandstone Formation.
c. Contour map of the base of the Mercia Mudstone Group
Only major faults are shown.

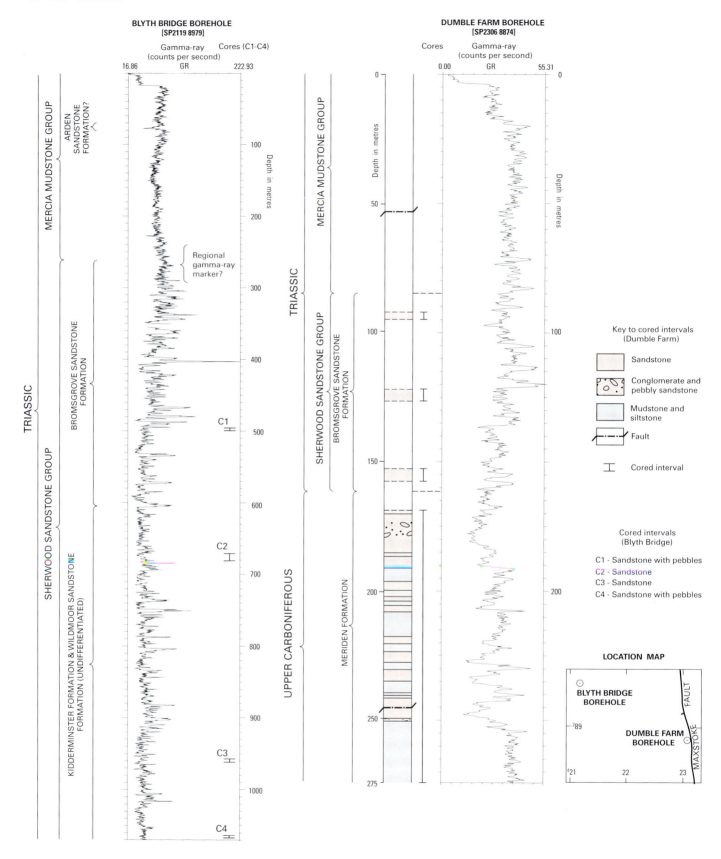

Figure 21 Gamma-ray logs and lithological log for cored parts of the Blythe Bridge and Dumble Farm boreholes.

Note different vertical and gamma-ray scales.

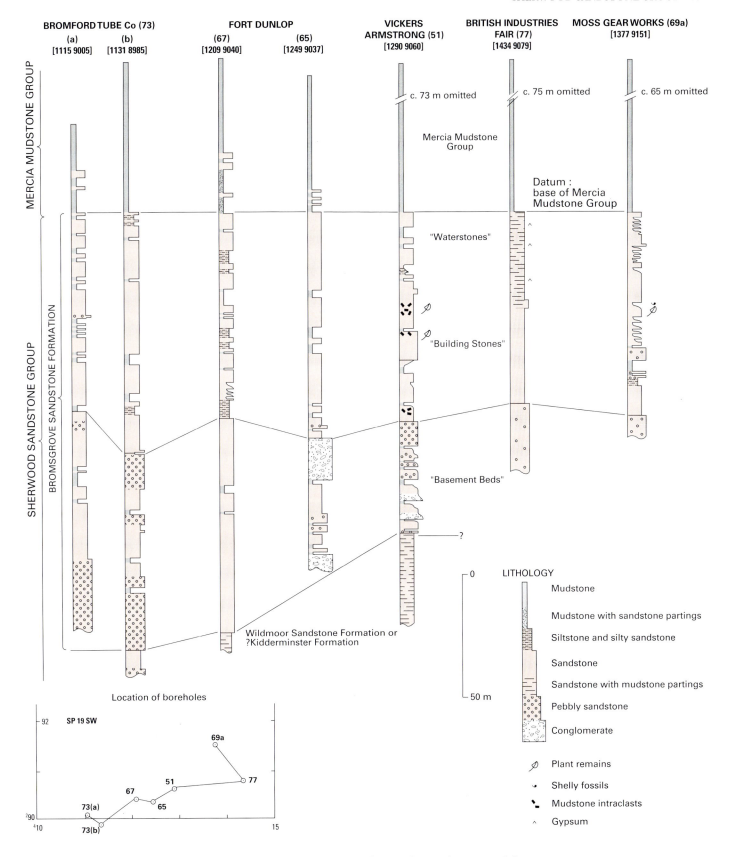

Figure 22   Borehole sections through the Bromsgrove Sandstone Formation east of the Birmingham Fault (after Piper, 1993a).

uplift and erosion in late Early Triassic times followed by a major change in provenance and sedimentation. The boundary of the Bromsgrove Sandstone with the overlying Mercia Mudstone Group is gradational and lies within an interbedded sandstone–mudstone sequence, about 6 m thick, termed the 'Passage Beds' (Butler and Lee, 1943); Wills (1976) considered these to be the basal beds the Mercia Mudstone Group. In this account, however, the base of that group is taken arbitrarily at the level where mudstone beds become predominant upwards.

The Bromsgrove Sandstone consists of red-brown, medium- to coarse-grained, subangular, arkosic sandstone, locally with scattered pebbles and beds of pebble conglomerate, together with subordinate thin beds of red mudstone and siltstone; caliche pellet conglomerate is common and the sandstone beds are generally calcite cemented. Well-rounded granules and small pebbles of quartz, quartzite and feldspar are common in the lower part of the formation.

Bedforms typically comprise trough cross-bedding in tabular or wedge-shaped cosets about 0.5 to 1.5 m thick; erosional scours are lined with quartz/feldspar pebbles and granules, and angular red mudstone intraclasts are common; foresets are generally graded. Reactivation surfaces and contorted lamination, resulting from dewatering, are present locally. Cross-bedding azimuths measured in the district indicate palaeoflow towards the north or north-east which accords with the regional trend (Warrington, 1968; Old et al., 1991). Repeated pebble-based, upward-fining cycles generally 2 to 5 m thick (Fitch et al., 1966; Warrington, 1968; Wills, 1976; Old et al., 1991), are particularly common in the lower part of the formation, and indicate pulsatory, high-velocity discharge which, together with the bedforms, indicates fluvial deposition in braided channels of a large river system. An increasing proportion of red, blocky mudstone beds in the upper part of the cycles higher in the formation indicates a change to deposition in a more mature, meandering river system with more extensive overbank deposits, or in shallow, ephemeral lakes. The formation shows an overall upward-fining trend, punctuated by pebble-based, upward-fining cycles (Figure 22). Up to eleven cycles ('cyclothems' I to XI) ranging in thickness from about 6 to 30 m, were recognised by Wills (1976, p.133, pl. VIIIA) in the best recorded wells in central Birmingham (Metro-Cammell, Southalls No. 4, Vickers Armstrong, Moss Gear Co. and J & E Sturge Ltd. Nos 1 and 2 boreholes (pp.113–114). These pulses represent major variations in stream velocity and sediment supply, possibly related to second or third order geomorphic cycles (Schumm, 1977; Wescott, 1993) which may have been controlled by climatic cyclicity.

The presence of well-rounded, quartz, quartzite and quartz-tourmaline 'schorl' extraclasts may indicate either rejuvenation of the Armorican source area which had supplied the conglomeratic Kidderminster Formation, or reworking of the latter since the Bromsgrove Sandstone is known to overstep the Kidderminster Formation in the district. However, the presence of unweathered feldspar grains in the Bromsgrove Sandstone indicates derivation

in part, at least from a separate alkali granite source, though possibly also from the Armorican source area which supplied the Kidderminster Formation (Fitch et al., 1966). The pebble suite also includes tuff, sandstone and siltstone, indicating more heterogeneous local sources, together with intraformational mudstone, siltstone and sandstone. Caliche pebbles ('cornstones'), present in the channel-lags, indicate high evapotranspiration rates and pedogenesis on the floodplain, and subsequent reworking by rivers scouring the floodplain during flood stages. Heavy mineral suites are dominated by staurolite, apatite, tourmaline and locally garnet (Fleet, 1923, 1929; Smithson, 1931), supporting a metasedimentary provenance. Further details of the petrology and sedimentology of the formation deduced from the area to the south are given in Old et al. (1991).

The Bromsgrove Sandstone has yielded a few fossils in the district; these include the branchiopod crustacean *Euestheria minuta*, the scorpion *Mesophonus*, fish scales, and plant remains (*Schizoneura*, *Voltzia* and *Yuccites vogesiacus*) from the Moss Gear Co. and Southalls boreholes (Butler and Lee, 1943); palynology samples from the J & E Sturge Ltd. No. 1 and Nechells Gasworks boreholes proved barren. A richer and more diverse floral and faunal assemblage has been recorded, together with microfloras, in the Droitwich and Redditch districts, around Bromsgrove (Old et al., 1991). This assemblage, principally from the middle (Finstall) member, comprises annelids (Ball, 1980), bivalves (Wills, 1910), arthropods, including scorpions (Wills, 1910, 1947) and crustaceans (*Euestheria minuta*), and vertebrates. The vertebrates are represented by remains of fish, including the lungfish *Ceratodus* (Wills, 1910), temnospondyl amphibians (Wills, 1910; Paton, 1974; Benton et al., 1994; Benton and Spencer, 1995) and reptiles, including a rhynchosaur, a rauisuchian, an archosaur, a lepidosaur and a nothosaur (Walker, 1969; Galton, 1985; Benton, 1990; Benton et al., 1994; Benton and Spencer, 1995). A similar vertebrate fauna is known from the formation in the Warwick district, to the south-east, (Old et al. 1987). Plant macrofossils from these districts include equisetalean pteridophytes and coniferalean gymnosperms (Arber, 1909; Wills, 1910, Robertson and McCallum, 1930). Microfloras (Warrington, 1970) reflect a more diverse flora of lycopsids, sphenopsids, pteropsids and gymnosperms, including conifers. The flora and fauna indicate continental, ephemeral floodplain environments with shallow lakes, which underwent periodic desiccation (lungfish) and were inhabited by semi-aquatic reptiles with marine affinity (lepidosaurs and nothosaurs), and other herbivorous and carnivorous reptiles (Old et al., 1991). The upper part of the formation (Sugarbrook Member of adjacent areas) yielded acritarchs and tasmanitid algae which, together with an upward decrease in the terrestrial biota, reflect a marine influence (Old et al., 1991).

Warrington et al. (1980) assigned an Early to Mid-Triassic (Scythian to late Anisian/early Ladinian) age to the Bromsgrove Sandstone Formation. However, following a reassessment of the microfloras and vertebrate faunas from the south Midlands (Benton et al., 1994; Barclay et al., 1997), it is now regarded as no younger than Anisian (early Mid-Triassic) Table 14; the age of the lower part of the Finstall Member, and the underlying Burcot Member of the Warwick area, is unproved.

There are few natural exposures of the Bromsgrove Sandstone in the district due to the drift cover and urbanisation; in central Birmingham sections in railway cuttings [0692 8674 and 0740 8680 to 0749 8675], at Sutton Coldfield [1217 9627], and on the Worcester and Birmingham canal near Calthorpe Fields [0527 8519 to 0538 8524] are given by Eastwood et al., (1925).

MERCIA MUDSTONE GROUP

The Mercia Mudstone Group crops out in a broad, down-faulted tract bounded to the north-west by the Birmingham Fault and to the east by the Western Boundary Fault of the Warwickshire Coalfield; much of the outcrop is obscured by drift deposits and urban development.

The maximum thickness, proved in boreholes in the east of the district, is about 365 m, but the full thickness in the Knowle Basin has not been proved in boreholes; geophysical evidence suggests that the formation may be 400 m thick in the central part of that basin. The only formal subdivision of the group, seen at outcrop in the district, is the Arden Sandstone Formation. This occurs about 320 m above the base of the group, is up to 10 m thick, and forms a mappable unit in the Coleshill–Packington area in the east of the district. The Blue Anchor Formation (formerly 'Tea Green Marls'), at the top of the group, is up to 12 m thick; it has been proved in boreholes in the south-east corner of the district but is not present at outcrop. To the south-west, in the Worcester Basin, the mudstone succession below the Arden Sandstone is termed the Eldersfield Mudstone Formation, and that between the Arden Sandstone and the Blue Anchor Formation is termed the Twyning Mudstone Formation (Barclay et al., 1997). Other formations have been recognised to the north-east, in the Nottingham area (Elliott, 1961; Charsley et al., 1990). The succession in the Birmingham district is, however, more readily comparable with that in the Worcester Basin.

Calcareous mudstone was formerly exploited for 'marling' in small pits over much of the outcrop, and for brick and tile clay in larger quarries such as Stonebridge Quarry [2045 8260]; the Arden Sandstone was worked locally for building stone (Matley, 1912).

The base of the Mercia Mudstone is gradational upwards from the Sherwood Sandstone Group through as much as 12 m of thin-bedded sandstone, siltstone and mudstone termed the 'Passage Beds' by some workers (Butler and Lee, 1943) and equivalent to the upper part of the 'Waterstones' of Hull (1869). Following Wills (1976) and Warrington et al. (1980), these beds are included here within the underlying Bromsgrove Sandstone, and the lower boundary of the Mercia Mudstone is arbitrarily taken at the level where mudstone predominates over sandstone. The upper boundary is taken at the base of the Westbury Formation (Penarth Group), where black mudstones with thin sandstones rest on the Blue Anchor Formation.

The group consists predominantly of red-brown mudstone with sparse grey-green beds. Grey-green reduction spots or haloes ('fish eyes') are present locally; some of these have cores of black, radioactive nodules, up to 10 mm diameter, surrounded by a grey-green reduction halo (Harrison et al., 1983). Veins and nodules of fibrous gypsum and anhydrite have been proved in boreholes, generally at depths greater than 20 m, but evaporites are absent from surface exposures due to dissolution in groundwater. The thickest development of these sulphates was proved in the Knowle Borehole [1883 7777], located immediately south of the district, where up to 3.5 m of coalescing gypsum/ anhydrite nodules and secondary fibrous gypsum veins with minor red mudstone were proved at about 51 m below the base of the Penarth Group (Old et al., 1991). Mudstones are generally blocky in texture, lack fine lamination, and consist of clay minerals, carbonates, quartz and iron oxides. Clay minerals, in this district, comprise illite (mica) and chlorite of detrital origin, together with subordinate exotic assemblages which include mixed layer smectite-chlorite and smectite-illite, sepiolite and corrensite (Dumbleton and West, 1966; Jeans, 1978; Taylor, 1983). Dolomite is the predominant carbonate, although calcite is also present (Jeans, 1978). Thin beds of hard, grey-green, carbonate-cemented siltstone and fine-grained sandstone ('skerry' of brickclay workers) occur at several levels. For example, the Weatheroak Sandstone (about 0.7 m thick) crops out in the south-west of the district and was proved in the J & E Sturge Ltd. boreholes [0587 7963; 0579 7960]; locally it forms mappable features. The Weatheroak Sandstone consists of pale grey, flaggy sandstone with green mudstone interbeds; copper mineralisation is present at the base of the sandstone, predominantly in the form of secondary malachite, but including native copper and cuprite (Old et al., 1991). Pseudomorphs after halite and desiccation cracks are common on the bedding surfaces of these beds, which are commonly ripple cross-laminated.

The Mercia Mudstone is poorly exposed in this district; it was formerly exposed in many 'marl' pits scattered across the outcrop, but most of these are now degraded. Parts of the succession below the Arden Sandstone were formerly exposed in brick-clay quarries at Holly Lane [120 914], Erdington and Jackson's Brick Pit, Stonebridge [2045 8260] (Plate 2) (Warrington, 1993a, b). At Holly Lane Brick Pit the succession (Figure 23) consists mostly of red-brown, structureless mudstone. The lower 15 m of the stratigraphically higher section at Jackson's Brick Pit, Stonebridge (Plate 2) is of similar lithology, but the upper 18 m comprises about 13 erosively based, upward-fining cycles (Figure 23). A typical cycle comprises a thin, grey-green, fine-grained, cross-bedded sandstone, commonly with convolute lamination and some ripples, passing up through pale red-brown, poorly laminated, silty mudstone with green mottles, to dark red-brown, structureless, compact mudstone with fewer green spots; the

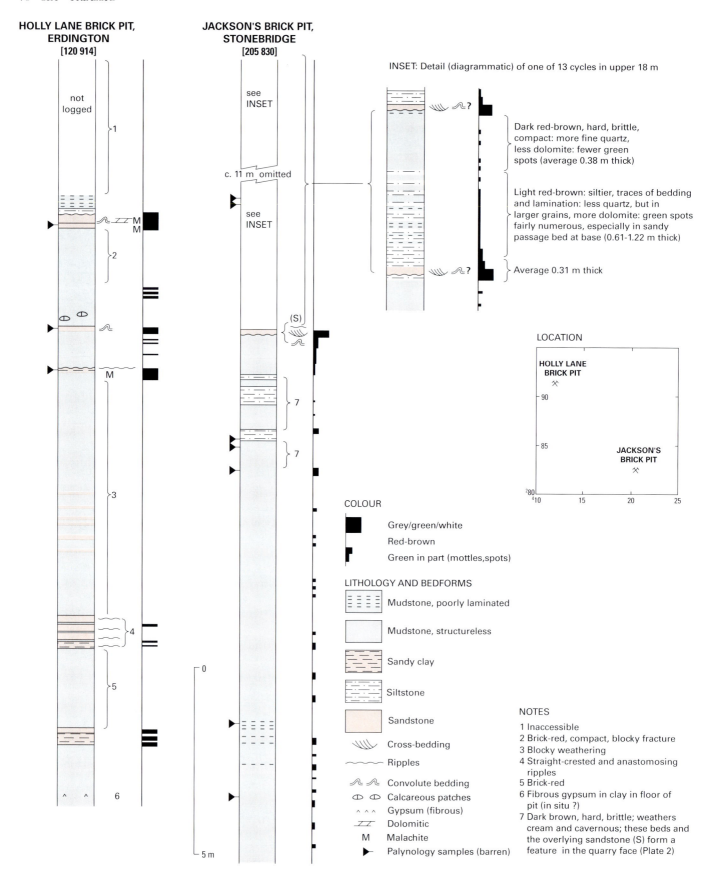

HOLLY LANE BRICK PIT, ERDINGTON [120 914]

JACKSON'S BRICK PIT, STONEBRIDGE [205 830]

INSET: Detail (diagrammatic) of one of 13 cycles in upper 18 m

Dark red-brown, hard, brittle, compact: more fine quartz, less dolomite: fewer green spots (average 0.38 m thick)

Light red-brown: siltier, traces of bedding and lamination: less quartz, but in larger grains, more dolomite: green spots fairly numerous, especially in sandy passage bed at base (0.61-1.22 m thick)

Average 0.31 m thick

LOCATION

HOLLY LANE BRICK PIT

JACKSON'S BRICK PIT

COLOUR

- Grey/green/white
- Red-brown
- Green in part (mottles,spots)

LITHOLOGY AND BEDFORMS

- Mudstone, poorly laminated
- Mudstone, structureless
- Sandy clay
- Siltstone
- Sandstone
- Cross-bedding
- Ripples
- Convolute bedding
- Calcareous patches
- Gypsum (fibrous)
- Dolomitic
- M Malachite
- ▶ Palynology samples (barren)

NOTES

1 Inaccessible
2 Brick-red, compact, blocky fracture
3 Blocky weathering
4 Straight-crested and anastomosing ripples
5 Brick-red
6 Fibrous gypsum in clay in floor of pit (in situ ?)
7 Dark brown, hard, brittle; weathers cream and cavernous; these beds and the overlying sandstone (S) form a feature in the quarry face (Plate 2)

Figure 23 Lithological logs of parts of the Mercia Mudstone Group exposed in Holly Lane Brick Pit and Jackson's Brick Pit (after Warrington, 1993a, b).

Table 15 Palynomorphs from the Arden Sandstone Formation.

Locality	Preparation numbers	Alisporites sp.	Alisporites parvus	Alisporites toralis	Brodispora striata	Camerosporites secatus	Cuneatisporites radialis	Cyathidites sp.	Cycadopites sp.	Duplicisporites granulatus	Duplicisporites verrucosus	Ellipsovelatisporites plicatus	Enzonalasporites vigens	Haberkornia gudati	Klausipollenites devolvens	Labiisporites sp.	Labiisporites granulatus	Ovalipollis pseudoalatus	Paravesicaspora sp.	Parvisaccites triassicus	Patinasporites densus	Podosporites sp.	Porcellispora longdonesis	Punctatisporites sp.	Ricciisporites umbonatus	Rimaesporites potoniei	Spiritisporites spirabilis	Triadispora plicata	Vallasporites ignacii	Verrucosisporites sp.	indeterminate bisaccate pollen	indeterminate circumpolles	indeterminate trilete spores	acanthomorph acritarch	Plaesiodictyon mosellanum
1	SAL. 2333	+		cf		+							?	+				+				+									+				
2	MPA 39302		+															+					+						?		+	+	?		
3	MPA 39303		cf		+	+	+	+	+				+	+	?	+	+	+	cf	+			+	+	+	+	+	+	?		+	+	?	?	+
4	MPA 39304		cf										?																		+				
	MPA 39305		cf			+				+	+	+			+	+	+	+	+		+			?			+	+	+		+	+	+		

1 Hampton in Arden [213 814]
2 Borehole 101R, Middle Bickenhill [20538 83135]
3 Borehole 102R, Middle Bickenhill [21297 83118]
4 Borehole 105R, Middle Bickenhill [21450 83127]

\+ occurrence
cf. comparable form present
? questionable occurrence
   Preparations curated in the palynology collections at BGS Keyworth

dolomite content and the size of the quartz grains decrease upwards through the cycle. The cycles record waning flow and probably represent discrete sheet-flood events over a low-gradient floodplain.

Arden Sandstone Formation

This formation was formerly known as the Arden Sandstone Member, (Warrington et al., 1980); it crops out in the east of the district where it forms a prominent escarpment in places, and has been quarried (Matley, 1912). The formation has a maximum thickness of about 10 m in the vicinity of Coleshill [200 890] and Hampton in Arden [200 810], but 5 m is more typical; it thins rapidly northwards and cannot be traced over the whole district. Boreholes indicate that it comprises pale green-grey, fine- to medium-grained, ripple cross-laminated and trough cross-bedded sandstone and siltstone, inter-bedded with pale grey mudstone and, locally, with red mudstone similar to the bulk of the group. In the Elmdon area, boreholes proved an upper leaf of green-grey siltstone and mudstone about 1.65 m thick, and a lower leaf of up to 3.9 m of grey-green mudstone, separated by 3 to 6 m of red-brown mudstone (Sumbler, 1982). In the Redditch district, to the south, palaeo-current measurements indicate flow towards the east (Old et al., 1991). The Arden Sandstone is probably equivalent to the Hollygate Skerry which lies at the top of the Edwalton Formation (Elliott, 1961) of the Nottingham area.

Blue Anchor Formation

Formerly the 'Tea-Green Marl', this formation consists of pale, grey-green, blocky mudstone and thin beds of siltstone with dolomite concretions. Calcite or dolomite forms up to 30 per cent of the red-bed lithofacies, probably reflecting an evaporitic depositional environment. Clay minerals include chlorite, corrensite, sepiolite and smectite (Jeans, 1978). Planar lamination, cross-lamination and desiccation cracks have been reported from adjacent areas (Taylor, 1983).

Age of the Mercia Mudstone Group

The Mercia Mudstone Group ranges in age from late Anisian to Rhaetian (Table 14). Fossils are rare in the district and comprise only palynomorphs. Palynology samples from the Saltley Baths Borehole and brickpits at King's Heath, Erdington and Stonebridge (Figure 23) proved barren but samples from the Arden Sandstone in boreholes near Middle Bickenhill, Stonebridge and a temporary section near Hampton in Arden station yielded miospores indicative of a late Carnian (Tuvalian) age (Table 15). Deposition in brackish water is indicated by the presence of the chlorococcalean alga *Plaesiodictyon mosellanum*, and a possible acanthomorph acritarch may suggest marine influence (Warrington and Ivimey-Cook, 1992). The miospore association reflects a vegetation dominated by gymnosperms, principally conifers; miospores from ferns and a bryophyte are also present, indicating damp terrestrial habitats. In the Redditch district, the Arden Sandstone has yielded comparable microfloras and a relatively diverse flora and fauna (Old et al., 1991) indicative of continental terrestrial and subaqueous brackish environments with marine connections, probably in an estuarine setting (Warrington and Ivimey-Cook, 1992).

The highest beds of the Mercia Mudstone Group, including the Blue Anchor Formation, proved in the Knowle Borehole located immediately to the south of the district, yielded sparse dinoflagellate cysts (*Rhaetogonyaulax rhaetica*) (Old et al., 1991), suggesting the onset of the Rhaetian transgression which established marine environments throughout much of England during deposition of the Penarth Group (Warrington and Ivimey-Cook, 1992).

Depositional environments

Red-bed mudstones and siltstones which comprise the greater part of the Mercia Mudstone lithofacies were deposited in continental environments ranging from shallow lakes to evaporitic inland sabkhas or playas fed by ephemeral streams on a low-gradient floodplain; sheet-flow and mudflow processes were probably important in transporting fine-grained siliciclastic sediment to the centre of the low-gradient basin. Finely laminated siltstone and silty mudstone were probably deposited from suspension in shallow lakes. Deposition by ephemeral streams of mudstone aggregates as sand-size bedload has also been suggested (Rust and Nanson, 1989; Talbot et al., 1994); the mudstone aggregates are thought to have been produced by erosion of floodplain sediments where the soils (vertisols) have undergone pedogenesis in an arid or semi-arid climate. The genesis of such aggregates seems to require the presence of a swelling clay, such as smectite, in the floodbasin; small amounts of smectite or mixed-layer smectitic clays have been reported from the Mercia Mudstone (Jeans, 1978; Taylor, 1983; Leslie et al., 1992) and a semi-arid to arid climate is inferred, so this process could have been important in the genesis of much of the structureless mudstone in the district. Mud aggregate textures have not been reported from the Mercia Mudstone Group in this district, but it seems likely that any original aggregate texture would be removed during compaction diagenesis. Some of the silt-size grains making up the blocky textured mudstones may have been deposited as loess by aeolian transport (Wills, 1976; Glennie and Evans, 1976; Arthurton, 1980; Talbot et al., 1994). The presence of desiccation cracks, evaporitic minerals (veins and nodules of gypsum) and pseudomorphs after halite indicate evaporation of surface waters and groundwater brines of continental and marine origin (Taylor, 1983) in a semi-arid to arid climate. This climatic regime is also supported by the ubiquitous red colour of the floodplain sediments, reflecting early diagenetic oxidation. Thin

sandstone and siltstone beds ('skerries'), with straight crested and anastomosing ripples and convolute bedding, were probably deposited from broad, shallow, ephemeral, braided streams by sheetflood processes. Their grey colour reflects more rapid deposition, probably during periods when the water-table was higher, preventing early diagenetic oxidation of the sediments. Similar conditions were more widespread, and persisted for longer, during deposition of the Arden Sandstone. The Arden Sandstone marks a major change in depositional environment over a large part of the basin; the widespread distribution, lithology and bedforms of this formation indicate deposition in a fluvial environment dominated by laterally migrating rivers. The presence of possible acritarchs and brackish water to marine faunal indicators (Old et al., 1991; Warrington and Ivimey-Cook, 1992) indicates connection with marine environments, possibly in estuarine situations.

Parallels have been drawn between the depositional environments of the Mercia Mudstone Group and arid coastal flats such as the Ranns of Kutch (Glennie and Evans, 1976; Arthurton, 1980) or arid to semi-arid, inland continental basins such as the eastern interior of Australia during the Quaternary (Rust and Nanson, 1989; Talbot et al., 1994). Both analogues may be valid for some parts of the Mercia Mudstone succession of the district, but not applicable to the full succession. Grey-green, dolomitic beds of the Blue Anchor Formation, at the top of the group, mark a change to less oxidising conditions on the floodplain, possibly as a result of a higher water-table and preservation of organic matter (algal) within the sediments. A marine, or estuarine influence is suggested by the presence of sparse dinoflagellate cysts in the Knowle Borehole (Old et al., 1991, fig. 13). In other areas (e.g. Somerset, Dorset), the presence of rhythmically interbedded mudstones and non-laminated fine-grained dolomites, locally with gypsum, suggests deposition in coastal sabkhas (Sellwood et al., 1970). The absence of sulphate evaporite minerals in this formation, and evidence from stable isotope studies in the Midlands (Taylor, 1983), indicate a mixed marine and continental water source, possibly reflecting the onset of the widespread Rhaetian transgression (Mayall, 1981).

PENARTH GROUP

The Penarth Group (Warrington et al., 1980) is equivalent to the 'Rhaetic' of earlier workers. It crops out on the southern margin of the district, near Copt Heath, and consists of about 12 m of dark grey mudstone with minor siltstone and limestone, locally with marine fossils. It represents continuation of the major transgression from the Tethys Ocean over a wide area of present-day, north-west Europe during Rhaetian time (Warrington and Ivimey-Cook, 1992). The group is subdivided into the Westbury Formation and the overlying Lilstock Formation. In this district the latter consists largely of the Cotham Member, although a thin limestone, proved in the Knowle Borehole (Old et al., 1991), may represent the overlying, carbonate-dominated Langport Member.

The group is poorly exposed in the district (Brodie, 1865; 1874) and most of the available information comes from the Knowle Borehole [1883 7777] (Institute of Geological Sciences, 1982, p.3; Old et al., 1991), located immediately south of the district.

Westbury Formation

In the Knowle Borehole, the Westbury Formation rests with a sharp base on the Blue Anchor Formation, and consists of 6.02 m of dark grey, fissile, mudstone with silty laminae, a thin bed (0.04 m) of limestone and a thin bed (0.15 m) of siltstone. The mudstone is pyritic and yields abundant bivalves, including *Rhaetavicula contorta*, *Lyriomyophoria postera*, *Chlamys valoniensis*, *Eotrapezium concentricum*, *E.* cf *germari*, and *Protocardia rhaetica*, together with the gastropod '*Natica*' *oppelii*, and fish scales such as *Gyrolepis alberti*. Cross-lamination, including climbing-ripples, has been recorded in areas to the south (Old et al., 1991). Similar lithologies and bedforms can be traced over a wide area of England (Ivimey-Cook and Powell, 1992; Benfield and Warrington, 1988), indicating the rapid establishment of relatively uniform, shallow-marine conditions over an extensive region. Palynomorph assemblages (Old et al., 1991, fig. 13) comprise rich associations of miospores and organic-walled microplankton which increase in diversity upwards through the formations; the microplankton, which consist largely of marine dinoflagellate cysts, dominate some of the associations. The microflora and macrofauna of the formation indicate a Rhaetian (late Late Triassic) age.

Lilstock Formation (Cotham Member)

In this district, and throughout the Midlands and in North Yorkshire (Raymond, 1955; Old et al., 1991); the Lilstock Formation is represented mostly by the Cotham Member, here about 6 m thick. The stratigraphically younger Langport Member is thicker in areas to the south (Williams and Whittaker, 1974; Old et al., 1987) where it has yielded conodonts (Swift, 1995a, b). This member may be represented in the Knowle Borehole by a 0.13 m thick, fissured limestone which shows two well-developed erosion surfaces and yields *Pteromya* cf. *tatei*. In the Knowle Borehole, the Cotham Member consists of 6.1 m of pale to medium-grey, calcareous silty mudstone with a few thin siltstone and limestone beds; analogous mudstone in north Yorkshire is rich in illite, chlorite and smectite clay minerals (Raymond, 1955). The boundary with the underlying Westbury Formation is gradational, with an upward change, over 0.4 m, from dark non-calcareous mudstone to pale grey calcareous mudstone in which slumped and contorted laminae are also common. Scattered bivalve fragments, including *Rhaetavicula contorta*, in these beds may have been derived from the Westbury Formation, since there is evidence of local erosion at the base of the Cotham Member.

Macrofossils are generally sparse, apart from abundant, small branchiopod crustaceans (*Euestheria minuta*), and the bivalves noted above, but miospores and marine microplankton, indicative of a Rhaetian age, are common (Old et al., 1991, fig. 13). Sedimentary structures and biota suggest that the Cotham Member was deposited in shallow, possibly brackish, marginal marine environments. The abundant and diverse miospore assemblages in the both the Cotham Member and the Westbury Formation contrast with the sparse assemblages in the Blue Anchor Formation, and suggest a rapid expansion and diversification of the contemporary flora, possibly in response to climate amelioration associated with the spread of epicontinental seas northwards from the Tethys Ocean in latest Triassic times (Warrington, 1981).

SEVEN

Jurassic

Rocks of early Jurassic age crop out in a small, fault-bounded outlier at Copt Heath, near Knowle, in the south-east of the district (Figure 1). They form part of the Lias Group (Powell, 1984), and are included in the Blue Lias Formation (Cope et al., 1980). The lowermost beds, 4 to 5 m thick in the Knowle Borehole [1883 7777] (Old et al., 1991, fig.4), below the first occurrence of the ammonite *Psiloceras planorbis*, are of Triassic age, but are described in this chapter, for convenience.

LIAS GROUP

The lithostratigraphy of the Lias Group used herein (Table 14), follows that established for the adjacent Redditch district (Old et al., 1991). Exposures are few and the succession is not known in detail, although where not obscured by drift deposits much of the outcrop is dotted with shallow pits dug for limestone, and many of the fields are covered with fragments of limestone and fissile mudstone. The Lias Group here is about 30 m thick, and consists predominantly of alternating beds of grey calcareous mudstone and subordinate thin limestone, deposited in a shallow epeiric sea. The 'Lias' transgression onlapped the London Platform, located to the south-east (Donovan et al., 1979), and although mudstone sedimentation was dominant, fluctuations in relative sea level, combined with reworking during storm events and relative low-stands, resulted in the deposition of bioclastic limestone. A list of fossils identified from the Lias Group in the district and the adjacent Redditch district is given in Old et al., 1991, appendix 5).

Blue Lias Formation

WILMCOTE LIMESTONE MEMBER

The basal member consists of 8 to 10 m of mudstone and, locally, siltstone, intercalated with thin beds of limestone. The limestones are generally fine-grained, argillaceous and flaggy weathering, and in the upper part of the unit are laminated; shelly fossils are locally common and some of the beds are bioclastic. The lowermost 2.5 m of the member, comprising dark grey mudstone with thin siltstone beds were penetrated in the Knowle Borehole; these Pre-planorbis Beds are considered to be latest Triassic in age (Old, 1982; Old et al., 1991). The Pre-planorbis Beds are estimated to be 4 to 5 m thick in this area (Old et al., 1991; fig. 14). Insect remains and a possible plant leaf have been collected from the Copt Heath area (Old et al., 1991, appendix 5). At Waterfield Farm [1904 7789], shale soil brash yielded the ammonite *Psiloceras planorbis*, and the bivalves *Modiolus minimus* and *Pteromya tatei*, indicating the *planorbis* Sub-biozone.

SALTFORD SHALE MEMBER

The Saltford Shale Member spans the *planorbis* and *liasicus* ammonite biozones (Old et al., 1991), and in this district consists of about 20 m of fissile and blocky mudstone with a few thin blue-grey argillaceous limestone beds and nodule bands. Limestone was formerly worked from three shafts at Copt Heath [1808 7808, 1814 7807 and 1805 7798], all of which were abandoned before 1857 (Brodie, 1865; Eastwood et al., 1925). Records of '*Ammonites planorbis*' from spoil around the shafts (Brodie, 1874) indicates a *planorbis* Sub-biozone age. An abundant fauna, present on some limestone bedding planes, includes the bivalves *Cardinia ovalis*, *Plagiostoma* sp., and *Pseudolimea hettangiensis*, the echinoid *Eodiadema minutum*, and ostracods. The Saltford Shale Member comprises the larger part of the Lias outcrop in the district, which differs from that of the previous survey in extending eastward in the vicinity of Barston [205 780] into an area formerly regarded as Arden Sandstone (Eastwood et al., 1925). The presence of the Lias Group there had, however, been recognised by Matley (1912), who published a faunal list, and was confirmed by a more recent survey in the Barston area (Old, 1987) which recorded fragments of limestone and grey-brown shale of probable *planorbis* or *liasicus* biozone age.

EIGHT

Quaternary (Drift)

Quaternary, superficial deposits (drift) cover about 75 per cent of the district. They consist predominantly of till and sandy till, glaciofluvial deposits (sand and gravel), and glaciolacustrine deposits (mostly laminated clay, silt and sand), most of which were probably deposited during the Anglian or Wolstonian glaciations (Kelly, 1964; Jones and Keen, 1993), and interglacial deposits (mostly peat, organic silt and clay) of Hoxnian age (Duigan, 1956; Kelly, 1964; Horton, 1974). The status of the Wolstonian (glacial) Stage in central England is uncertain (Shotton and West, 1969; Sumbler, 1983; Rose, 1987, 1991; Horton, 1989; Shotton, 1989; Gibbard, 1991; Maddy et al., 1994); the chronology of the glacial and interglacial deposits (Table 16) in the district is discussed, below. Glacigenic sediments (Table 17) were deposited as subglacial and englacial facies such as till (diamicton) and as lens- or sheet-like bodies of sand and gravel. Meltwater streams emanating from the ice sheets

Table 16 Stratigraphy of the Quaternary (drift) deposits of the district.

System/Series		Stage (cold stages shown in italics)	Deposit	Approximate age (where known) in years BP[1]
QUATERNARY	HOLOCENE	FLANDRIAN	Alluvium (silt, clay, sand, gravel and sparse peat) Peat and lacustrine deposits Landslip Head First river terrace deposits	12 000 to present day
	PLEISTOCENE (part)	*Devensian*	Second river terrace deposits (Hams Hall Terrace) ?Till in Willenhall/Darlaston area	12 000–110 000
		IPSWICHIAN	? not present; periglacial solifluction (Head)?	110 000–130 000
		Wolstonian (?) (see text for discussion)	? Glaciofluvial deposits (sand and gravel) ?Glaciolacustrine deposits (clay, silt and sand) ? Till and sandy till	130 000–352 000
		HOXNIAN	Peat, organic-rich silt and clay (Nechells, Quinton and Grimstock Hill)	352 000–423 000
		Anglian (= 'Wolston Series' of Sumbler, 1983)	Glaciofluvial terrace deposits (clayey gravel) Glaciofluvial deposits (sand and gravel) Glaciolacustrine deposits (clay, silt and sand) Till and sandy till Head (Quinton)	> 423 000

1 After Bowen (1994) and Wessex Archaeology (1996).

Table 17
Outline of the classification of glacial and associated interglacial deposits of the district.

Lithology	Deposit	Landform	Environment of deposition
Till and sandy till (clay diamicton, and sandy diamicton)	Till and sandy till (boulder clay)	Degraded undulating terrain; also present in palaeovalleys	Subglacial (lodgement till) and supraglacial (melt-out and flow tills)
Sand and gravel (clayey, in part)	Glaciofluvial deposits (undifferentiated)	Broad, degraded undulating spreads, elongate ridges (eskers); also present in palaeovalleys	Supraglacial, englacial and subglacial streams; outwash plains; proglacial braided rivers, and ice-marginal streams
	Glaciofluvial terrace deposits	Flat-topped spreads	Proglacial rivers or ice-marginal streams
Clay, silt and sand (mostly laminated)	Glaciolacustrine deposits	Broad, flat spreads; also present in palaeovalleys	Shallow, impounded lakes fed by proglacial streams
Organic-rich silt and clay, and peat	Interglacial deposits	Present in palaeovalleys (proved in boreholes)	Colonisation of shallow lakes and depressions by plants during temperate stages; deposition of organic matter in shallow lakes

deposited glaciofluvial sand and gravel as proglacial outwash. In some areas, meltwater was impeded, resulting in the formation of lakes in which laminated clay, silt and sand were deposited, locally rich in organic matter; peat was deposited during low lake-level stands, associated with temperate interglacial stages.

Glacial deposits covered much of the district prior to erosion during Devensian and Flandrian times (Table 16), when much of the glacigenic sediments were re-deposited as alluvial sand and gravel. These are now preserved as the river terraces and alluvium of the Rivers Tame, Rea, Cole and Blythe, and their tributaries. Down-slope movement (solifluction and gelifluction) of weathered bedrock and superficial deposits resulted, locally, in deposition of head. Drift-free areas occur in places, such as on the higher ground in central Birmingham, the South Staffordshire Coalfield (particularly the area west of the watershed which corresponds approximately to the trace of the Russell's Hall Fault), the Warwickshire Coalfield, and the flanks of the river valleys which dissect the largely drift-covered terrain of the central part of the district.

The district has a special significance for Quaternary geology in the English Midlands since it includes the classic Hoxnian interglacial succession, proved in boreholes in the Nechells Green area [08 88] (Duigan, 1956; Kelly, 1964; Horton, 1974), and Quinton area [99 84] (Horton, 1974; Horton, 1989), and the well-defined, sediment-filled palaeovalleys (buried channels) (Figures 24, 25). Acquisition of site investigation boreholes during this resurvey has allowed a revision of the Quaternary

sequence, and of the form and nature of the deposits infilling the palaeovalleys.

GLACIAL DEPOSITS

Till and sandy till (boulder clay)

Till and sandy till are present over much of the south-west of the district, and small patches also occur in the centre and east. The boundary between till and sandy till is gradational and is not shown on the map; the deposits represent lithological end-members of a clay-dominated and a sand-dominated till (diamicton). The matrix of the tills consists of red, orange-brown, brown and yellow clay and sandy clay, with lenses and admixtures of clayey sand; thicker beds and lenses of clayey sand are present locally, but these are generally too small to be distinguished separately on the map. Clasts generally comprise well-rounded pebbles and cobbles. The most common clasts are pebbles and cobbles of red-brown quartzite, 'milky' (vein) quartz and sandstone reworked from the conglomerates of the Triassic Kidderminster Formation. Sparse igneous and metamorphic pebbles are also present, and these too, were probably derived from this formation. Subordinate, subrounded clasts of coal, ironstone, grey mudstone and sandstone are derived from the Coal Measures, and red mudstone is derived from the Mercia Mudstone Group. The lithology of the till and sandy till closely resembles that of the 'red Triassic-derived boulder clay — Thrussington Till' of the Redditch district (Old et

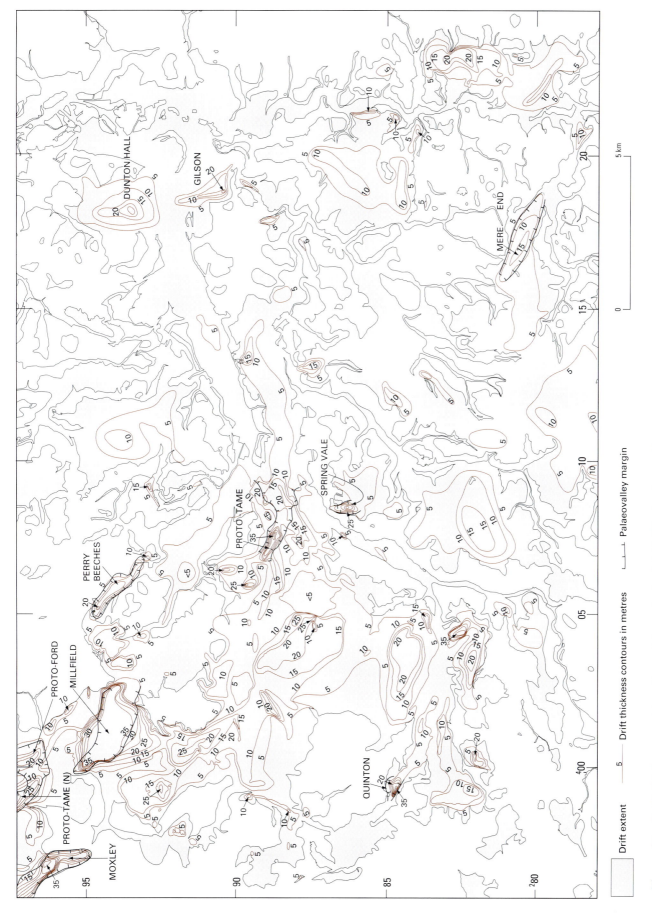

Figure 24 Drift thickness and location of the principal drift-filled palaeovalleys.

Drift extent

5 —— Drift thickness contours in metres

⊢——⊣ Palaeovalley margin

0   5 km

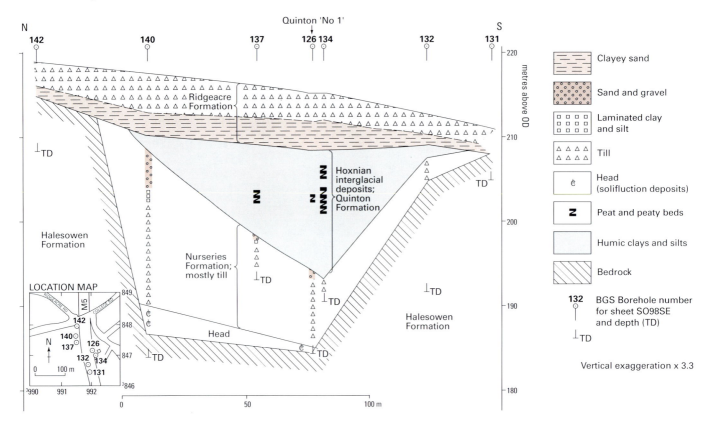

Figure 25   Cross-section through the Quinton palaeovalley; formation names and information after Horton (1989).

al., 1991) and Warwick district (Sumbler, 1983). In addition, large erratic boulders, up to 3 m in diameter, composed of rhyolite and andesite have been described from a number of localities in the district (Martin, 1891), but the greatest concentration occurs at Bournville [04 81] (Lapworth, 1913) and Selly Oak [03 82] (Laurie, 1926); these erratics are thought to be derived from the Ordovician (Arenig) volcanic rocks of North Wales.

In a study of Quaternary deposits proved in site investigation boreholes in central Birmingham, Kelly (1964) described four stratigraphically distinct tills. In descending sequence these are the Edgbaston Till, Brookfields Till, Snow Hill Till, and Aston Till, and are separated by unnamed units of glaciofluvial sand and gravel or glacial lake deposits. However, the stratigraphical superposition and lateral persistence of these tills cannot be demonstrated in the records of recently acquired shallow boreholes, and it is doubtful that they represent a laterally persistent, 'layer-cake' stratigraphy. The lateral persistence of Kelly's named tills was also questioned by Horton (1974), who examined numerous site investigation boreholes in north Birmingham, including the Aston Expressway, and concluded that there is a complex sequence of interdigitating glacigenic lithologies throughout the Quaternary sequence of the district.

Some of the named tills, such as the Aston Till, described in boreholes by Kelly (1964, p.551), consist of clayey pebbly sand and sand, which are lithologically similar to some of the clayey glaciofluvial deposits (sand and gravel) described here. Some of the clayey sands with diamict textures, interpreted as till by Kelly (1964), may represent localised pockets or lenses of poorly sorted glaciofluvial sand and gravel, or soliflucted blocks of till incorporated in the glaciofluvial deposits.

Till, up to 18 m thick, comprising red-brown clay with pebbles of sandstone, Triassic quartzite and igneous rocks derived from North Wales, Scotland or the Lake District, is present in the north-west of the district, north of an approximate east–west line from Moxley to King's Hill [Northing 96] (Hamblin et al., 1991). The till overlies, and is therefore younger than, the deposits in the Moxley palaeovalley, which are assumed to be Anglian or Hoxnian in age (see below). The till has been ascribed to the Devensian glaciation (Tables 16, 18), which transported 'Irish Sea Till' to the north-west of the district (Martin, 1891; Eastwood et al., 1925; Hamblin et al., 1991); the southern limit of this ice sheet is thought to be represented by the line noted above. However, it has not been possible to map the boundary between this 'upper or Devensian' till, on lithological or morphological characteristics, from the till present farther south, since both deposits have a high percentage of 'Triassic' quartzite pebbles. Furthermore, the age of the 'upper' till is uncertain; it could be ascribed to either the glacial Wolstonian or Devensian stages (see discussion, below).

Table 18 Outline of the alternative theories regarding the age of the Anglian to Devensian glacial succession of the district.

Stage	Willenhall/Darlaston area (NE of district)	Quinton area (after Horton, 1989)	Nechells area (after Duigan, 1956; Kelly, 1964)	North and west Birmingham (alternative hypotheses (a) and (b), after Horton, 1974, 1989)	
Devensian	Till in Willenhall/Darlaston area	Not present			
Ipswichian	Possible solifluction deposits				
Wolstonian	Older drift sequence (Anglian or Wolstonian) in palaeovalleys	Till Sand and gravel (Ridgeacre Fm.)	Glaciofluvial sand and gravel	(a) Glaciofluvial sand and gravel, till, glaciolacustrine deposits	Not present ?
Hoxnian		Interglacial deposits (Quinton Fm.)	Interglacial deposits	Not present ?	
Anglian		Till (Nurseries Fm.) and head	Glaciofluvial sand and gravel		(b) Glaciofluvial sand and gravel, till, glaciolacustrine deposits

The 'Earlswood Till Complex' (Old, 1983) consists of an intimate association, up to 9 m thick, of red-brown till, laminated clay, silt, sand and gravel, and is interbedded with, or passes laterally into, glaciofluvial deposits (Old, 1983; Old et al., 1991). The 'complex' was mapped at the 1:10 000 scale (SP 07 SW) around Solihull Lodge [094 787], where laminated clays were locally worked for brickclay (Old, 1983), but it has not been distinguished separately from other till on the 1:50 000-scale map. Old et al., (1991) interpret the 'complex' as having been deposited close to the front of the Western (Irish Sea) Ice Sheet (Anglian or possibly Wolstonian in age) as it advanced southwards into a predominantly lacustrine environment.

Glaciofluvial deposits, undifferentiated (sand and gravel)

Glaciofluvial deposits are present, at the surface, over much of the central and east of the district, and have been worked for aggregate at numerous sites in the district (see Chapter Two). Similar deposits are also present in sediment-filled palaeovalleys such as the Proto-Tame, Proto-Rea, and Moxley 'buried channels' (Figure 24), including the sites at Quinton and Nechells (Figures 25, 26), where they are interstratified with glaciolacustrine deposits and interglacial deposits.

Glaciofluvial deposits consist largely of poorly sorted, red, orange and yellow sand, clayey sand, pebbly sand, and gravel. Sand and gravel is generally orange-red, red-brown or yellow in colour, and ranges from poorly to well graded. Grain-size analysis of deposits in the south-east of the district (Cannell, 1982; Cannell and Crofts, 1984) indicate the following mean gradings: fines 12 per cent, sand 63 per cent and gravel 25 per cent, although there is considerable vertical and lateral variation from pebble-free sand to gravel. The sand fraction is largely composed of fine- to medium-grained, angular to subrounded quartzite and quartz. Most of the grains are coated with iron oxides which impart a red or yellow-brown colour to the deposits (Cannell, 1982). Sands are usually trough cross-bedded, with pebble-lags lining erosive surfaces; gravel lenses are common. Contorted bedding has been described from many exposures (Chilcott, 1922; Eastwood et al., 1925, pp.110–111; Powell, 1993a); this may be due to collapse following melting of stagnant ice, post-depositional slumping or dewatering, or deformation by over-riding ice sheets (Horton, 1974).

Granule and pebble clasts are mostly rounded to well-rounded and consist largely of Triassic-derived quartzite, quartz and sandstone, with coal, mudstone and ironstone fragments derived from upper Carboniferous rocks, including the Coal Measures, occurring locally. Other clasts include Silurian limestone and siltstone and sparse exotic pebbles of quartz porphyry and igneous rocks; Shotton (1977, p.18) recorded granite, rhyolite, ignimbrite, andesite and tuff, thought to be derived from

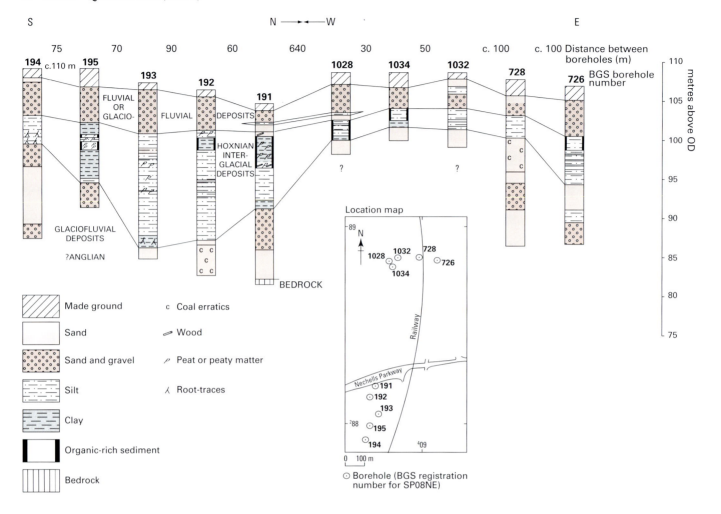

Figure 26   Correlation of boreholes in the Nechells Green area, including the Hoxnian interglacial site (boreholes 191 to 195; Duigan, 1956; Kelly, 1964).

North Wales, the Lake District or Scotland. Sections in the glaciofluvial sand and gravel between Key Hill and Hockley Hill [around 059 883] revealed rounded and subrounded clasts, including felsite and Llandovery sandstone, thought to be derived from North Wales (Crosskey, 1882).

Glaciofluvial sand and gravel, together with interbedded glaciolacustrine and interglacial deposits, were proved in boreholes to be up to 36 m thick within the sediment-filled palaeovalley (Proto-Tame) that trends parallel to Hockley Brook [077 885] (Figure 27). Sand and gravel was formerly worked in many shallow pits in the district such as Bustleholm Sand Pit [015 945] which proved about 19 m of red sand with lenses of quartzite pebbles and detrital coal, and thin beds of clay; bedding is highly contorted (Chilcott, 1922). Away from the palaeovalleys, the deposit generally ranges in thickness from 5 to 10 m. In the workings around Cornets End [23 81], the deposit is typically 10 m thick, but farther south along the Blythe Valley [20 80] it is 17 m thick, and generally comprises gravel overlain by sand (Cannell, 1982). Cross-bedding in this area indicates a palaeoflow to the south (Sumbler, 1982).

In the Nechells area (Figure 26), sands and gravels of presumed glaciofluvial origin are present, above and below interglacial deposits which are regarded as Hoxnian in age (Duigan, 1956; Kelly, 1964). The stratigraphical position of the Hoxnian interglacial deposits suggests that the underlying glaciofluvial sand and gravel is Anglian in age; whether the overlying glaciofluvial deposits at this site, and intercalated tills described from adjacent areas (Kelly, 1964), are all post-Hoxnian age is open to debate (see below). Horton (1974) suggests that the named tills (Kelly, 1964) and glaciofluvial deposits (sand and gravel) around Soho Road [05 89] may, in part, predate the Hoxnian interglacial deposits at Nechells, and, therefore, are also Anglian in age.

The age of the glaciofluvial sand and gravel overlying the Hoxnian interglacial deposits at the Nechells (Figure 26) and Quinton sites (Figure 25) (Horton, 1974; Horton, 1989) is less certain. At the Nechells site, these deposits were regarded as resting unconformably on the underlying interglacial deposits (Kelly, 1964), on account of the temperate to early glacial part of the interglacial profile being missing. However, the boundary may be merely erosional; such erosional contacts are common in

Figure 27 Rockhead contour map for part of central Birmingham. Contours at 5 m intervals.

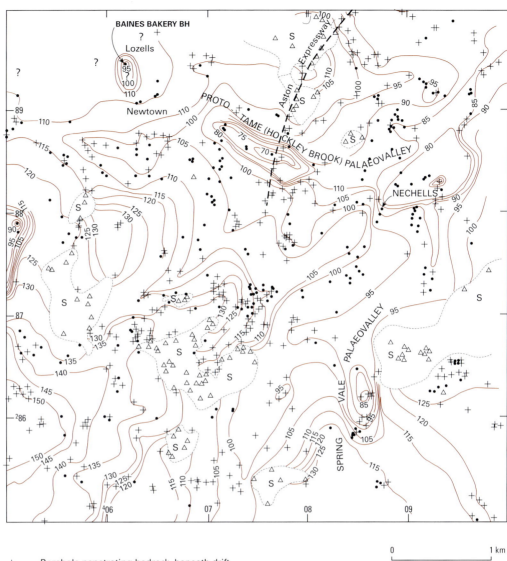

+	Borehole penetrating bedrock, beneath drift
•	Borehole terminating in drift
△	Borehole penetrating bedrock at, or near, surface (less than 1.5m drift thickness)
S	Areas of thin drift or bedrock outcrop

fluvial and glaciofluvial sequences. The uppermost sand and gravel deposits may, therefore, represent glaciofluvial outwash of post-Hoxnian age; that is, it was deposited during the subsequent Wolstonian Cold Stage (Shotton, 1953; Shotton, 1989). Alternatively, it may be fluvial in origin, representing sedimentation during the later part of the Hoxnian Interglacial Stage.

Glaciofluvial terrace deposits

Spreads of 'fluvioglacial' gravel were mapped at an elevation of about 100 m above OD along the eastern margin of the Blythe valley, between Outwoods [23 84] and Four Oaks [24 80] (Sumbler, 1982). They lie at a higher topographical level than the undifferentiated glaciofluvial deposits described above, and are shown on the map as 'Glaciofluvial terrace deposits'. They overlap the margins of the 'Glaciofluvial deposits' (undifferentiated), and consist of up to 10 m of poorly sorted 'clayey' sand and gravel with till-like lenses (Sumbler, 1982; Cannell, 1982); clasts consist of Triassic-derived quartzite and quartz with some sandstone, mudstone and sparse igneous clasts. At Cornets End Gravel Pit [234 811], the terrace deposit is extensively cryoturbated, with ice-wedge structures extending to 1.3 m into the underlying undifferentiated glaciofluvial deposits (Sumbler, 1982). The glaciofluvial terrace deposits are thought to have been laid down by melt-waters flowing from an ice-front located to the north (Cannell, 1982). The age of these deposits is uncertain; they may be Wolstonian or Devensian in age since they overlie undifferentiated glaciofluvial deposits of presumed Anglian age.

Glaciolacustrine deposits

Glaciolacustrine deposits consisting of grey-red and brown clay and laminated silty clay, together with laminae and thin beds of fine-grained sand, have been proved at surface, and in boreholes, at a number of sites in the district. These deposits are commonly associated with thick drift sequences infilling palaeovalleys, such as those at Moxley [971 952], Quinton [992 847], Nechells [087 881], Millfield [024 941] and California [036 834]; extensive deposits also occur in the east of the district near Dunton [190 932], Gilson [190 907] and Brickfield Farm [195 864]. Small spreads are also present adjacent to the Blythe Valley in the south-east of the district.

In the Nechells Green area (Figures 24, 26), the glaciolacustrine deposits form part of the Nechells glacigenic sequence infilling the Proto-Tame (Hockley Brook) palaeovalley (Duigan, 1956; Kelly, 1964; Horton, 1974). At this site, the 'Late-glacial lake series' of Kelly (1964) is about 2 m thick, and similar deposits about 3.5 m thick were proved in recently acquired borehole records. Glaciolacustrine deposits are locally overlain by peaty or humus-rich Hoxnian Interglacial deposits, and generally underlain by glaciofluvial sand and gravel, of presumed Anglian age (Figure 26). The re-survey (Powell, 1993b) has demonstrated that the glaciolacustrine and associated organic interglacial deposits have a much wider distribution than previously thought; they can be traced, at depth, from the Nechells site [087 881], to an area about 900 m farther north, where these deposits have been proved in boreholes [088 898].

The fine lamination present in the glaciolacustrine clay, silty clay and fine-grained sand (Kelly, 1964) suggests that they were deposited, from suspension, as varves in a proglacial lake. Similar laminated glaciolacustrine lake clays, proved in motorway boreholes in the Birmingham area, have been shown by Horton (1974) to be interbedded with glaciofluvial sand and gravel, and these deposits, together, are generally underlain by till. The 'motorway sequences' suggest a depositional environment of glaciofluvial outwash and locally ponded lakes in a proglacial setting. In the Nechells area, however, the presence of an almost complete Hoxnian interglacial profile (see below), overlying the glaciolacustrine deposits indicates ponding-up of the Proto-Tame (Hockley Brook) depression, which was partly infilled by late Anglian times, and subsequent shallowing and silting up of a lake, followed by deposition of humic deposits during the Hoxnian interglacial. The boundary of the glaciolacustrine deposits and overlying interglacial deposits is gradational and was taken arbitrarily, above the varved clays (Kelly, 1964; base of bed 3a). There is a marked upward increase in organic content and the abundance of microspores above this level.

At the Quinton site, the glaciolacustrine deposits (silt and clay) do not appear at outcrop, but were proved in boreholes (e.g. Quinton No. 1 (126), and adjacent boreholes; Figure 25). They form part of the Quinton Formation of Horton (1974; 1989), which also comprises glaciofluvial sand and interglacial humic clays and peat. A few kilometres farther east, at California [041 836],

glaciolacustrine deposits occur on either side of Bourn Brook valley (Pickering, 1957), and on the north-east side of Wood Brook valley (Glover, 1990b). Hereabouts, they consist of sporadically laminated, brown and orange silt, silty clay and sandy clay; the 'rubbery' clays were formerly worked for brick clay. Small pebbles present in the clays were interpreted as dropstones (Pickering, 1957; Glover, 1990b). The glaciolacustrine deposits in this area appear to have been deposited in a depression located along the present-day Bourn Brook valley and extending to Weoley Castle [02 82] (Glover, 1990b). Towards the east and north, these deposits are intercalated with glaciofluvial sand and gravel. The glaciolacustrine deposits at California were deposited at an elevation of about 160 m above OD, considerably lower than those at the Quinton site (about 200 m above OD). If the deposits are coeval, then the difference in elevation suggests that lakes were formed at various levels on the pre-existing 'Anglian' land surface.

Glaciolacustrine deposits within the Millfield (Proto-Tame) palaeovalley (Horton, 1974; Waters, 1991b) are up to 32 m thick, and include, in addition to the lithologies noted above, an abundance of locally derived coal detritus. The elevation of the deposits, hereabouts, ranges from about 68 m above OD to about 110 m above OD.

The Moxley palaeovalley (Figure 24) is infilled by an upward-fining sequence which, west of Willenhall [965 986], comprises glaciolacustrine silts and clays intercalated with sand and gravelly clay, overlying glaciofluvial sand and gravel and till. Farther south, at Moxley [971 956], glaciolacustrine laminated silts and clays overlie fine-grained, well-sorted sands and gravels, up to 35 m thick. To the south of Moxley, the sand occurs at surface and was formerly worked in the Moxley Sand Pits [972 956].

Glaciolacustrine deposits are also present in a north-trending palaeovalley (Figure 24), near Dunton [190 931] (Waters, 1993). Brown (1980) considered this palaeovalley to represent the eastern continuation of the Proto-Tame, proved in central Birmingham (e.g. Nechells) (Kelly, 1964). The glaciolacustrine deposits overlie till, and are interbedded with glaciofluvial sand and gravel; they comprise yellow, orange and red-brown laminated clayey sand and silt with thin beds of pebbly clay and gravel. The pebbly clays may represent dropstone deposits released by melting ice sheets. Small tracts of glaciolacustrine deposits in the south-east corner of the district may represent the northernmost deposits of the Meer End Channel (Old et al., 1991) which fed into glacial Lake Harrison, postulated by Shotton (1953) to have been impounded by ice, to the south of the district.

Interglacial deposits, undifferentiated

Humic clay, silt and peat, interpreted as deposits of the Hoxnian interglacial have been proved in boreholes at two well-known sites in the district, namely Nechells (Duigan, 1956; Kelly, 1964; Shotton and Osborne, 1965) and Quinton (Horton, 1974; 1989) (Figure 25, 26). These deposits have not been proved at surface during this survey. Similar Hoxnian interglacial deposits were proved at Grimstock Hill (Brown, 1980) and at another site located to the north-west of the district, at Trysull (SO 84 94), south-west of Wolverhampton (Morgan, 1973).

NECHELLS

Boreholes in the Nechells Green area [08 88] (Figure 26; Nechells Redevelopment and adjacent boreholes 191–195) proved up to 10 m of humic silty clay and clay with finely comminuted organic detritus, woody peat, and humic fine- to medium-grained sand with coarse organic debris (Kelly, 1964). Organic-rich deposits were temporarily exposed during site development in the 1960s. These deposits were termed the 'Interglacial Series', and on the basis of their pollen and insect content were assigned to the Hoxnian interglacial stage (Duigan, 1956; Kelly, 1964). Site investigation boreholes (Figure 26), located about 900 m to the north (Powell, 1993b), proved similar deposits at about the same elevation (about 95 m above OD).

Fine, varve-like lamination in these deposits, is less pronounced than in the underlying glaciolacustrine deposits (Kelly, 1964). Peat is present as comminuted wood fragments together with drifted branches and logs. The vegetation, incorporated in the lake mud and sand, records an ameliorating climate; surrounding forest tree cover included *Betula*, *Pinus*, *Quercus*, *Alnus* and *Taxa*, representing, in part, temperate deciduous forest (Duigan, 1956; Kelly, 1964). These are followed by mostly coniferous forest (*Picea*, *Abies* and *Pinus*) and grassy heathland flora, indicative of a return to a colder, drier climate and suggesting a possible transition to a subsequent glacial stage (possibly the Wolstonian Stage), although part of cold climate profile is missing (Kelly, 1964). The flora and fauna of the Hoxnian interglacial deposits at Nechells are similar to records from interglacial deposits at Quinton (Horton, 1974), Grimstock Hill (Brown, 1980) and at Trysull (Morgan, 1973).

The interglacial deposits are underlain by glaciofluvial deposits (sand and gravel) of presumed Anglian age, and are overlain by glaciofluvial or fluvial sand and gravel, of unknown age. These sands and gravels may have been deposited during the Hoxnian Stage or, perhaps, during the subsequent Wolstonian or later Devensian glaciations (see discussion, below).

QUINTON

At the Quinton site, located close to the M5 motorway in west Birmingham, interglacial deposits, proved in two boreholes [9921 8471 and 9954 8456], are underlain by till and overlain, in upward sequence, by glaciofluvial sand and till (Horton, 1974) (Figure 25; boreholes 126, 134 and 137). The interglacial deposits consist of humic clay, silt and sand with layers of plant debris and thin beds of peat (Quinton Formation of Horton), which yielded a palynological record considered to extend through the Hoxnian to the early Wolstonian (Herbert Smith, cited in Horton, 1989). These deposits are considered to be coeval with the Hoxnian interglacial deposits of the Nechells area. However, the top of the interglacial deposits at Quinton lies at about 210 m above OD

compared to about 95 m above OD at Nechells (located only 10 km to the north-east). This suggests local impounding of lakes during the Hoxnian interglacial on an irregular topographical surface of Anglian deposits.

GRIMSTOCK HILL

Hoxnian interglacial deposits were proved in boreholes at Grimstock Hill, near Gilson [192 902] (Brown, 1980). They consist of brown silt and peat, about 3 m thick, and yielded a flora interpreted by Brown (1980) as indicative of the Hoxnian interglacial.

GLACIAL PALAEOVALLEYS (SEDIMENT-FILLED CHANNELS)

The preglacial (Anglian) topography of the district, determined from thousands of borehole records, is highly irregular (Figures 24, 27). The district is characterised by a number of incised palaeovalleys (buried valleys) infilled with glacigenic sediments, and in places, by interglacial deposits. A number of these palaeovalleys have been described in previous studies: Proto-Tame (including Nechells) (Pickering, 1957; Kelly, 1964; Horton, 1974), Proto-Tame-north/Millfield (Horton, 1974), Proto-Tame at Dunton (Brown, 1980), Moxley (Jukes, 1859; Eastwood et al., 1925; et al., 1982), Proto-Rea (Pickering, 1957) and Quinton (Horton, 1974; 1989). However, acquisition of borehole records during this re-survey has allowed the nature, form and infill of these palaeovalleys to be refined and, in addition, a number of new palaeovalleys have been discovered. The new borehole data have shown that the earlier interpretation of the palaeovalleys as interconnected valleys with a smooth, gradational thalweg profile (Pickering, 1957; Horton 1974, fig. 27) is incorrect. The rock-head contours (e.g. Figure 27) show that the general form of the palaeovalleys (Figure 24) is similar to the hull of a boat, that is, deeply incised towards the centre and shallowing at either end. This suggests that the valleys were eroded, either by subglacial scour or subglacial streams which eroded the bedrock under the confining pressure of overlying ice (Boulton and Hindmarsh, 1987), or perhaps as 'jökulhaup' plunge pools which cut back into bedrock and unroofed overlying ice, at the ice-sheet margin (Ehlers and Linke, 1989; Wingfield, 1989).

The palaeovalleys are infilled by a variety of glacigenic and interglacial deposits, representing deposition in subglacial and englacial, glaciofluvial and glaciolacustrine environments. The lithology of the fill is highly variable, both within and between individual palaeovalleys. However, glaciolacustrine and glaciofluvial deposits are usually present, and stratigraphically important Hoxnian interglacial deposits are present at Quinton, Nechells and at Grimstock Hill. The most significant palaeovalleys are described briefly, below.

Proto-Tame (Hockley Brook) palaeovalley (including the Nechells area)

The most extensive palaeovalley in the district is the Proto-Tame (Hockley Brook) palaeovalley, which trends east–west approximately parallel with Hockley Brook (Figure 27). This was formerly termed the Proto-Tame and considered to represent the former course of the River Tame which flows roughly parallel, but a few kilometres to the north. The form and extent of the Proto-Tame was first outlined by Pickering (1957), and subsequently revised by both Kelly (1964) and Horton (1975). More recently acquired borehole records have allowed further revision of the shape, extent and nature of the fill (Figures 26 and 27). Bedrock contours show that, contrary to previous models, the palaeovalley trends west-east and not north-west–south-east (Horton, 1974, fig. 7). This indicates that the palaeovalley might not represent an early bedrock channel related to the Tame, but a more deeply incised palaeovalley of Hockley Brook, which is now a minor watercourse located approximately along the axis of the buried palaeovalley. Furthermore, the former interpretations show the Proto-Tame crossing the present-day River Rea at right angles in the vicinity of the Nechells site, and extending farther eastward [09 88] (Horton 1974, fig. 7). The revised bedrock contour map (Figure 27), however, shows that the palaeovalley trends north-east from the Nechells site, that is, along the trend of the present-day River Rea, which has its confluence with the Tame about 1 km to the north-east [09 89].

The most deeply incised part of the palaeovalley is in the vicinity of the Aston Expressway [07 88] (Horton 1974, fig. 8) where the elevation of the bedrock floor is about 65 m above OD, and the glaciofluvial fill (mostly glaciofluvial sand and gravel) is about 36 m thick. Hereabouts, the palaeovalley has a V-shaped cross-section, and in plan shows a distinctive south-east-trending, 'boat-shaped' form (Figure 27), suggesting deep scouring of the bedrock within the Aston area [07 88]. Erosional scouring appears to have been controlled, in part, by hardness of the bedrock; erosion is most intense within the softer Wildmoor Sandstone, and the palaeovalley floor rises in the vicinity of the subcrop base of the harder Bromsgrove Sandstone. The harder unit seems to have caused the palaeovalley to become both shallower and broader, and for its course to be deflected to the north-east where it links up with the Rea valley, north of the Nechells interglacial site.

To the west of Newtown [06 88], the palaeovalley bifurcates 'upstream' (Figure 27). The north-west-trending bifurcation, formerly extended in this direction as the Proto-Tame (Horton, 1974, fig.7), appears, on present borehole evidence, to be 'blind'. In the Lozells area [06 89], it is difficult to determine the thickness of the drift in driller's borehole logs because the weathered Wildmoor Sandstone bedrock is easily misinterpreted as 'sandy drift'. In this area, the supposed location of the Proto-Tame, most of the boreholes penetrated up to 10 m of sand and gravel without reaching bedrock. However, a single, old record at Baines Bakery [0619 8948], records 'ballast, coal, sand and clay' to 37 m depth — whether all this thickness represents drift is uncertain. If it is all drift, then there is some justification for extending a branch of the palaeovalley to this location. However, the presence of bedrock, at crop, about 500 m to the north-west of this location, along the supposed line

of the Proto-Tame indicates that the palaeovalley cannot be traced farther in this general direction.

The elevation of the floor of the palaeovalley in this area is between 105 and 110 m above OD. This compares with a markedly lower elevation of about 65 m above OD at the centre of the 'boat-shaped' depression, located about 1.5 km to the east, giving a drop in elevation of the floor of the palaeovalley of about 40 m in 1.5 km. The rockhead contour map (Figure 27) highlights the highly irregular topography of the preglacial bedrock surface along the thalweg of the palaeovalley. The irregular, but locally deeply scoured, form of the bedrock surface suggests that it was formed either as a result of glacial scouring, perhaps by valley glaciers, or by subglacial fluvial erosion which produced the V-shaped profile in the vicinity of the Aston Expressway (Horton, 1974).

Borehole records (Figure 24) indicate that the north-west-trending Proto-Tame depression of previous authors (Horton, 1974, fig. 7; Pickering, 1957; Kelly, 1964) is not present between Park Farm [03 91] and Soho Road [05 89] (Waters, 1991b; Powell, 1993b). The palaeovalley located near Friar Park [00 95] is interpreted as an extension of the Millfield Channel (Waters, 1991b) which does not extend southward beyond Newton [02 93]. It seems unlikely that there was an earlier 'Tame' valley orientated parallel to, and located south of, the present-day River Tame in this area. Furthermore, the Proto-Tame (Hockley Brook) palaeovalley between Nechells Green [08 88], in the east, and Gib Heath (Soho Road) [05 89], in the west, represents an eastward-draining bedrock depression located south of the Tame, and roughly along the line of the former course of Hockley Brook (Figure 27).

The palaeovalley fill is lithologically highly variable but, as noted by Horton (1974), consists predominantly of sand, and sand and gravel, with subordinate amounts of clay and sandy clay with gravel. At the axis of the palaeovalley the drift fill is about 35 m thick (Figure 27); near the Nechells interglacial site, it is about 23 m thick, and in small 'bulls eye' depressions located north-west and south-east of the River Rea, it is 21 m and 31 m thick, respectively. The latter figure includes recent river alluvium. Horton (1974) suggested that the palaeovalley fill proved in boreholes along the Aston Expressway was equivalent to the glaciofluvial outwash deposits (possibly of Anglian age) *underlying* the Hoxnian interglacial deposits. This is borne out by borehole data which show that there is little evidence for the presence of the named, laterally persistent tills, overlying the interglacial deposit, as figured by Kelly (1964). Sandy clay with pebbles (possible till) in the Aston area cannot be traced as marker beds, as proposed by Kelly (1964), and appear to be small pockets within the general glaciofluvial sequence. The age and genesis of the sand and gravel overlying the interglacial deposits at both the classic Nechells site (see above) and the borehole locality to the north [088 888] (Figure 26) is uncertain (see discussion, below).

Proto-Tame (north)/Millfield palaeovalley

The north-west- to west-trending Proto-Tame (north)/Millfield palaeovalley (Figure 24) extends from Mesty Croft

[00 95] through Bustleholm [014 945] to Newtown [025 935]. In this area, the Proto-Tame (Pickering, 1957) and the Millfield Channel (Waters, 1991b) are synonymous. However, Pickering suggested that the palaeovalley could be traced to the south-east, beneath Park Farm [035 919] and Handsworth [04 90] to link up with the Proto-Thame in the Nechells area (Horton, 1974, fig. 7). Recent borehole acquisitions show that the palaeovalley terminates abruptly to the west of Newton, and cannot be traced to the south or east (Waters, 1991b).

The Millfield palaeovalley has an asymmetrical, broadly U-shaped cross-section, with a steeper northern margin; its base is cut to about 70 m above OD. The palaeovalley fill is up to 35 m thick, and comprises till at the base, overlain by a complex succession dominated by glaciolacustrine clay, silt and sand, intercalated with thin, laterally impersistent beds of sand and gravel and till (Waters, 1991b). The glaciolacustrine deposits were interpreted by Horton (1974) to have been laid down in a proglacial lake occupying the palaeovalley. The palaeo-valley fill is overlain by a thick succession of glaciofluvial sand and gravel and thin till. Hoxnian interglacial deposits have not been recorded from this palaeovalley, so the relative ages of tills, above and below the glaciola-custrine deposits, is uncertain.

Proto-Tame and Proto-Ford palaeovalley

The southern extensions of these palaeovalleys (Figure 24) are located in the north-west of the district near Pleck [99 96] and Bescott [00 96]. The Proto-Tame palaeovalley, in this area, underlies the course of the present-day River Tame. The palaeovalley is cut to about 80 m above OD, is steep sided, and has an irregular floor which rises towards the south-east. The fill consists of up to 30 m of till overlain by complex intercalations of glaciolacustrine silt and clay, and poorly sorted glaciofluvial gravelly clay, sand and gravel.

At its southern end, near Bescott [00 96], the Proto-Tame palaeovalley joins up with the north-east-trending Proto-Ford palaeovalley (Figure 24), which underlies the course of the present-day Ford Brook. The palaeovalley is infilled by up to 18 m of pebbly sand and gravel, but glaciolacustrine deposits have not been recorded.

Till and glaciofluvial deposits which blanket much of the north-east of the district outside of the palaeovalleys have been equated with a younger glaciation (possibly Wolstonian or Devensian in age) typified by deposits in the Walsall and Wolverhampton areas (Hamblin et al., 1992), By analogy with the palaeovalleys at Quinton and at Nechells, the lowermost till within the Proto-Tame and Proto-Ford palaeovalleys is Anglian in age. However, the age of the upper till, glaciolacustrine deposits and glaciofluvial deposits within the palaeovalley is uncertain.

Moxley palaeovalley

The Moxley palaeovalley is located in the north-west of the district (Figure 24), and is orientated parallel to the Proto-Tame. The palaeovalley is steep sided, and the base is cut to about 100 m above OD; at Moxley [971 956] the fill comprises up to 40 m of glaciofluvial sandy gravel and fine-grained, well-sorted sand, locally overlain by glaciola-custrine laminated silts and clays (Eastwood et al., 1925, p.110). The palaeovalley ends 'blind' [972 948] to the north-east of Gospel Oak.

By analogy with the age of the deposits within the palaeovalley at Nechells (see above), the Moxley deposits are probably Anglian to Hoxnian in age.

Quinton palaeovalley

The palaeovalley at Quinton [99 84] (Figure 24) is small in area, but boreholes proved about 30 m of drift deposits (Figure 25), with the base of the palaeovalley at about 185 m above OD (Horton, 1974; Horton, 1989). The deposits, formally defined by Horton at this locality, comprise, in upward sequence, a basal head deposit (about 4 m thick, comprising fragments of Upper Carboniferous bedrock and exotic pebbles), a lower till (Nurseries Formation) (up to 7 m thick, consisting of brown sandy till with erratic pebbles of locally derived Upper Carboniferous bedrock, dolerite, 'Triassic debris' and 'Silurian' limestone), interglacial and glaciolacus-trine sand, silt and clay with abundant organic matter and peat (Quinton Formation) (up to 15 m thick) and the Ridgeacre Formation (about 8 m thick, consisting of a lower unit of sand with clay partings, and an upper red-brown sandy till unit).

The lower till (Nurseries Formation) is interpreted as Anglian in age (Table 18) since it is overlain by organic-rich interglacial deposits which yielded a palynological record indicative of the late Anglian (cold) to Hoxnian (temperate), and possibly early Wolstonian (cold) stages (Herbert Smith, cited in Horton, 1989). The uppermost sand and till unit (Ridgeacre Formation), if glacigenic in origin (Horton, 1989), must have been deposited during a younger glaciation, possibly during the Wolstonian or even the Devensian stages (Table 18). However, Glover (1990a) suggested that the uppermost unit might represent head (a gelifluction or solifluction deposit), so that a later glaciation need not be inferred at this site (see discussion, below).

Other palaeovalleys

In addition to the palaeovalleys noted above, incised palaeovalleys (Figure 24) occur at **Spring Vale** [08 86] (base at about 80 m above OD), where up to 29 m of glaciofluvial deposits have been proved, by boreholes (Figure 27) (Powell, 1993b); at **Perry Beeches** [05 94 to 06 93] where boreholes proved up to 32 m of drift compris-ing, in upward sequence, till, glaciolacustrine laminated silt and clay and glaciofluvial sand and gravel (Waters and Powell, 1994); and at **Dunton** [185 933] where the base of the palaeovalley is at about 70 m above OD, and is infilled with sandy till, overlain by intercalated glaciolacustrine silt and clayey sand with thin beds of pebbly clay and gravel, and glaciofluvial sand and gravels (Brown, 1980; Waters, 1993). An Anglian age was proposed for the fill at Dunton (Brown, 1980), based on correlation with similar glaciola-custrine/glaciofluvial deposits which are overlain by

Hoxnian interglacial peat, at a nearby site at Grimstock Hill [192 902]. In the south-east of the district, Sumbler (1982) proved a drift-filled palaeovalley adjacent, along part of its length, to the course of the River Blythe. This palaeovalley appears to extend to the south of the district as the **Mere End Channel** (Old et al., 1991). The fill consists largely of glaciofluvial sand and gravel, with subordinate glacio-lacustrine deposits and till.

CHRONOLOGY OF THE GLACIAL DEPOSITS IN THE BIRMINGHAM AREA

The age of the glacigenic deposits in the Birmingham area has been the subject of much debate. However, elucidation of the chronostratigraphy of these sediments is hampered by the paucity of natural exposures due to urbanisation, a reliance on information gained from site investigation boreholes, poorly defined geomorphological features, and a paucity of stratigraphical marker beds.

Hoxnian interglacial deposits at Quinton (Figure 25) are underlain and overlain by a diamicton interpreted as a till (Horton, 1974; 1989) or possibly a redeposited till (Glover 1990a), and those at Nechells (Figure 26) are underlain and overlain by glaciofluvial sand and gravel (Duigan, 1956; Kelly, 1964; Powell, 1993b). Although these sites are only about 10 km apart, there is a difference of elevation of about 100 m above OD. This suggests that the Hoxnian interglacial and associated glaciolacustrine sediments were deposited in locally ponded depressions (within pre-existing palaeovalleys), on an irregular 'Anglian' topography. The underlying glacigenic sediments, comprising till at Quinton and glaciofluvial sand and gravel at Nechells, are considered intuitively to have been deposited during the Anglian glaciation, based on their stratigraphical position. By inference, the varied glacigenic sediments infilling the other palaeovalleys in the district (see above) are also considered to be of a similar age. With the exception of the Hoxnian interglacial peat deposits at Grimstock Hill (Brown, 1980), they lack dateable, organic-rich Hoxnian sediments, but commonly comprise till or glaciofluvial sand and gravel associated with glaciolacustrine deposits. The latter represent quiet-water sedimentation, possibly coeval with the Hoxnian glaciolacustrine/interglacial sedimentation, although there is no direct evidence for this.

The outstanding debate centres on the age and genesis of the post-Hoxnian deposits at the Nechells, Quinton and Grimstock Hill sites, and the age of the till and glaciofluvial deposits found, at surface, over much of the district. Do they represent deposits of the later Wolstonian Glacial Stage, as their superposition suggests (Brown, 1980; Shotton, 1989), or can they be explained by local factors? Regarding the Quinton site (Figure 25), the post-Hoxnian till (Ridgeacre Formation) could be interpreted as a solifluction deposit, possibly derived from same till which underlies the interglacial deposits, but which also emerges from the palaeovalley to blanket the higher ground in the area. If this were the case, these deposits need not be representative of a later glaciation. The uppermost sand and gravel at the Nechells site

(Figure 25) could be interpreted as predominantly fluvial in origin, and deposited during the later part of the Hoxnian interglacial.

Although Kelly (1964) figured cross-sections showing 'named', laterally persistent tills interbedded with post-Hoxnian sand and gravels in the area adjacent to Nechells, the present borehole data suggest that beds of glacial diamicton cannot be correlated laterally over the area. Furthermore, Horton (1974; 1989) suggested that similar tills proved in motorway boreholes in north Birmingham, cannot be regarded as mappable marker horizons since they are both laterally impersistent and show a complex interdigitating relationship with glaciolacustrine and glaciofluvial deposits.

Glaciofluvial deposits overlying Hoxnian interglacial deposits at Grimstock Hill, and found at a similar elevation to the Nechells Site (88 to 100 m above OD), include pebbles with vertical long-axes. This fabric has been interpreted variously as resulting from solifluction (Brown, 1980) or cryoturbation (Waters, 1993); the latter process could have taken place during the Wolstonian or Devensian cold stages.

In the South Yardley/Ackock's Green area [10 80], till generally overlies glaciofluvial sand and gravel, suggesting a younger age for the till (Piper, 1993c). However, the lithological distinction between these deposits is not clear cut, and in many areas the two lithologies are interbedded and vary rapidly both vertically and laterally. This supports the earlier hypothesis of Horton (1974) who suggested that complexly interbedded till, glaciolacustrine and glaciofluvial deposits, proved in north Birmingham, and equivalent to the named tills of Kelly (1964), could be interpreted as Anglian in age (i.e. coeval with the deposits *underlying* the Hoxnian interglacial deposits), so that a Wolstonian glaciation need not be invoked for these deposits (see debate in Keen, 1989; Jones and Keen, 1993).

In the central Midlands, south of the district, deposits formerly considered to represent the Wolstonian (glacial) Stage (Shotton, 1953) have been re-interpreted as being equivalent to the deposits of the type Anglian Stage (Perrin et al., 1979; Sumbler, 1983; Rose, 1987; Rose, 1991; Maddy et al., 1994). Whether the sand and gravel deposits in the district *overlying* the Hoxnian interglacial deposits at Nechells were deposited during the subsequent Wolstonian glaciation, or even the later Devensian glaciation, is open to debate. The complex Quaternary glacigenic deposits in the district could, conceivably, be the product of a rapidly fluctuating glacial (cold) and interglacial/interstadial (temperate) climate which caused pulsatory advance and stagnation of a single, major ice-sheet from Anglian to Hoxnian times. The peats and humic deposits characterised by temperate floras and faunas, and associated glaciolacustrine sediments, at the Quinton, Nechells and Grimstock Hill sites, are local in extent and, furthermore, the Quinton site occupies a much higher elevation. If the organic deposits are coeval, then the lakes within the palaeovalleys must have developed on an irregular topography on the underlying glacigenic deposits. The temperate floras and faunas may reflect a brief amelioration of the climate (that is a late Anglian to early Hoxnian interstadial) of similar ecological zonation to, but earlier than, the type Hoxnian interglacial at Hoxne in Suffolk. The organic-rich/lacustrine sites in Birmingham may have been over-ridden by a subsequent southward advance of the ice sheet in late Anglian times. This hypothesis, favoured by the authors, can explain the complex interdigitation of glacigenic, glaciolacustrine and interglacial deposits in the district, and mitigates against the advance of a later Wolstonian ice sheet in the district. An alternative hypothesis attributes the majority of the glacigenic deposits, outside the palaeovalleys, to the Wolstonian glaciation (Shotton, 1989).

Another controversial problem, in the district, is the extent of the deposits of the subsequent Devensian ice sheet which covered much of the area to the north of the district. Clay-rich till with a high proportion of large 'Lake District' and 'Scottish' erratics is common in the Wolver-hampton and Cannock areas, to the north and north-west of the district; the southward limit of these deposits lies approximately along a line drawn from north Dudley, through Walsall towards Sutton Coldfield (Martin, 1891; Eastwood et al., 1925, fig. 1). This is regarded as the limit of the Irish Sea ice-sheet, of presumed Devensian age. This ice-sheet would have overridden the earlier Quater-nary deposits of Anglian and, possibly Wolstonian age. Although the line marking the southern limit of the Devensian ice sheet is thought to cross the north-west margin of the district, its precise position is not clear, largely as a result of the difficulty of distinguishing tills which are characterised by similar, locally derived erratics in a poorly exposed urban area. Furthermore, site investigation borehole records generally do not record the lithology of erratics, which would allow the provenance of the deposits to be deduced. A line marking the southern limit of till in Bilston–Moxley area [96 96], and extending onto the Dudley district to the west, may correspond with the southern limit of the Devensian ice sheet (Figure 24). As this line is traced eastward it crosses the Moxley, Proto-Tame (north) and Proto-Ford palaeovalleys which are infilled with glacigenic deposits of presumed Anglian age (glacial palaeovalleys). Thus, in the Moxley–Darlaston area, it is not possible to distinguish later Devensian glacigenic deposits from the earlier deposits of Anglian or Wolstonian age.

In summary, the conflicting interpretations of the complex drift sequences, based mostly on poorly logged site investigation records, the lack of exposure in the urban area and the broadly similar erratics in the drift lithologies, make it difficult to subdivide the drift deposits in the district on a chronostratigraphical or lithostratigraphical basis. The general geographical and temporal relationships are shown in Tables 16 and 18, and Figure 28.

POSTGLACIAL DEPOSITS

Postglacial deposits in the district (Table 16) are gener-ally the products of erosion and deposition during the latter stages of, and subsequent to, the last glaciation (Devensian). Erosion of the earlier glacigenic deposits probably took place under periglacial conditions when the Devensian ice front was located close to the north-west corner of the district. However, deposits of the second river terrace (Hams Hall Terrace) may predate this event and may be late Ipswichian or early Devensian in age (see below). Postglacial deposits, which consist predominantly of alluvial and lacustrine sediments, are mainly derived from the glacial deposits, with variable amounts sourced from the local bedrock. In addition, there are small areas of peat, head and landslip. Man made deposits (made ground) are present throughout much of the urban area and in areas of major construc-tion and engineering; the natural topography has also been influenced by man in the form of excavations (e.g. cuttings, pits and quarries), some of which are shown on the map. Areas of man-made ground are shown more extensively on the component 1:10 000 scale maps.

River terrace deposits

These are most common along the course of the Rivers Tame, Cole, Rea and Blythe. The topographically higher second river terrace is associated with the Rivers Tame and Blythe. In the absence of chronological data (e.g. fossils; artifacts), the chronology of the River Terrace deposits is tentatively based on correlation with other areas, such as the Avon Basin (Tomlinson, 1925; 1935) and Severn Basin (Wills, 1938; Shotton, 1953; 1954). However, the numbering of the terraces in the district (Tame/Trent Basin) has only local significance. River Terrace deposits of the Blythe valley range from gravel to very clayey sand and gravel with a mean grading of fines 9 per cent, sand 58 per cent and gravel 33 per cent (Cannell, 1982).

Second river terrace deposits

The deposits of the topographically higher second river terrace of the River Tame, locally referred to as the Hams Hall Terrace (Shotton, 1954), forms small spreads, up to about 6 m thick, though generally about 2 to 4 m thick. They consist of locally clayey, pebbly sand and gravel; clasts generally comprise well-rounded quartz and quartzite, derived from the older drift and local Triassic bedrock, together with sparse weathered granite and quartz porphyry. Cross-bedding is locally present. The terrace was not recognised during the previous survey; these areas were shown either as 'boulder clay' or 'undif-ferentiated drift'. In places the sand and gravel is exten-sively cryoturbated (deformation resulting from periglacial freeze–thaw); many of the larger pebbles are re-orientated with their long axes vertical. In some ploughed fields the admixture of bedrock fragments (Mercia Mudstone) gives the appearance of a till with pockets of gravel. Between Hams Hall [21 92] and Bodymoor Heath [20 95] much of the deposit has been removed by quarrying.

Typically, the top of the second river terrace of the River Tame lies at an elevation of about 7 to 9 m above the level of the present-day floodplain (alluvium). The elevation (in metres above OD) of the top of the terrace

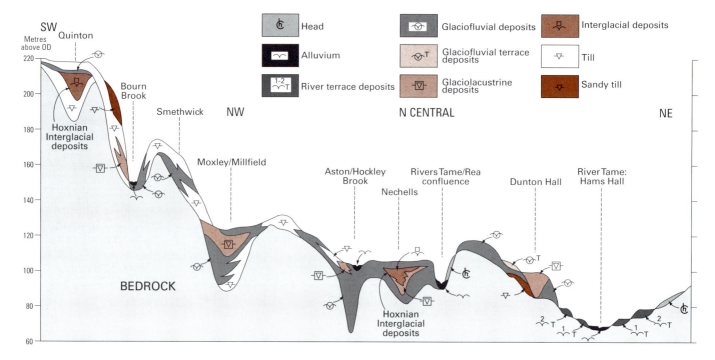

Figure 28   Schematic cross-section showing the relationships of the Quaternary deposits.
Exaggerated vertical scale.

ranges from 110 in the west, near Perry Barr [06 92], 95 to 90 in the Erdington and Stechford areas [11 90], 90 in the Nether Whitacre area [215 911] to 75 to 71 in the Middleton area [21 96]. The 'downstream' elevation of the agradational terrace, therefore, ranges from about 110 m above OD in the west to about 70 m above OD in the east; this represents a gradient of about 1 in 500.

The second river terrace of the River Tame has been equated with Avon No.4 Terrace and is thought to be either late Ipswichian or early Devensian in age (Shotton, 1953; 1954; Brown, 1980), or post Hoxnian to pre-Ipswichian (Bridgland et al., 1989) in age; however, no direct age criteria are present in the district. The second terrace of the River Cole may be of the same age. Cryoturbation, noted above, may have occurred during the Devensian glaciation.

First river terrace deposits

Deposits of the first river terrace form broad, intermittent tracts adjacent to the present-day floodplain of the larger rivers in the district, principally the Rivers Tame, Cole, Blythe and Rea. However, extensive quarrying for aggregates, urban landscaping and associated diversion of the water courses into man-made culverts makes it difficult to determine the extent and natural boundaries of the deposit. Thus, the boundaries of the first river terrace deposit, in much of the urban area, are largely taken from the previous survey, although in some areas the form of the terraces was barely visible even at that time (Eastwood et al., 1925, p.119).

The first river terrace forms a topographical feature about 1 to 4.5 m above the present-day floodplain (allu-

vium), and is equivalent to the 'Low Terrace' of Shotton (1954) and Brown (1980). It is generally 1 to 2 m thick, though exceptionally it reaches 7.8 m thick in the Hamstead area [022 936]. The lithology is similar to the second river terrace (Hams Hall Terrace), being predominantly composed of massive-bedded, and cross-bedded gravel and pebbly sand with well-rounded pebbles of Triassic-derived quartz and quartzite. Brown (1980) recorded coarsely bedded, horizontally laminated sand and gravel with thinner interbeds of coarse-grained sand in workings near Middleton Hall, located to the north of the district. Cryoturbation is present locally. Black silty clay and peat associated with pebbly sand and gravel have been recorded at surface and in boreholes in the near Hatchford Brook [151 852]; these deposits were attributed to the first terrace of the River Cole and its tributary Hatchford Brook (Eastwood et al., 1925).

The topographical level (in metres above OD) of the first river terrace deposit of the River Tame varies from about 130 in north-west of the district (Darlaston), to about 95 in central Birmingham, 86 near Erdington, 77 near Nether Whitacre and 74 in the north-east around Water Orton.

The first river terrace deposits of the River Tame were considered by Brown (1980) to be coeval with the Avon No. 1 Terrace (Shotton, 1954); Brown (1980) postulated a Mid-Devensian age, based on dating of peat underlying the deposits in the Middleton area. Analysis of pollen and insect remains, including beetles, collected from organic deposits preserved at the base of the first river terrace at Middleton Hall [SP 1940 9820], just to the north of the district on the Lichfield Sheet indicate an arctic climate (Brown, 1980), consistent with the local presence of cryoturbation noted above. Radiocarbon dating of these deposits yielded a ^{14}C age of 35 560 + 870 - 790 years BP (Brown, 1980), equivalent to the Upton Warren Interstadial Complex (Coope, 1975). The first river terrace deposits of the Rea may be of a similar age.

Along the valley of the River Blythe, the first river terrace deposits lie at about 3 m above the present-day floodplain, and were worked for sand and gravel near Ryton End [214 796].

Alluvium

The larger streams and rivers such as the Tame, Cole, Rea, Blythe and Hockley Brook are flanked by alluvial floodplains. The broadest spread of alluvium, about 3 km wide, occurs at the confluence of the Rivers Tame, Cole and Blythe near Blythe End [21 91].

In the urban areas, the alluvium is largely obscured by made ground commonly associated with the construction of flood defences, landscaping and the diversion of the rivers and streams into man-made culverts. Consequently, the natural boundaries of the alluvium in much of the urban area cannot be traced, and are largely based on the previous survey (Eastwood et al., 1925). This is especially so along the course of the River Tame [10 89 to 18 91], Hockley Brook and the central parts of the Rea valley, where it is difficult to distinguish terraced made ground from the natural alluvial floodplain. From Hams Hall [21 91] northwards to Kingsbury [20 96], much of the former alluvium has been extracted, in places, together with the first river terrace deposits, in aggregate quarries. Bedrock (Mercia Mudstone Group) was exposed at the base of many of these workings, and the voids have been either partly backfilled, or flooded.

The alluvium typically comprises an upper layer of brown clay or silt, up to about 3 m thick, underlain by several metres of pebbly sand and gravel. Thin lenses of peat may be present, locally. However, considerable lithological variation is found, both laterally and vertically, even in the same river valley, so that any of the aforementioned lithologies may be present at surface. Pebble types are similar to those of the second and first terrace deposits, from which they were largely derived during incision of the rivers during the Flandrian. In the Nechells area, the Tame/Rea alluvial tract overlies the Proto-Tame (Hockley Brook) palaeovalley fill, and although up to 30 m of 'sandy drift' has been proved in boreholes [093 883], only the upper part is alluvium.

The previous survey recorded pebbly sand and gravel at surface in a number of temporary sections, but the most significant record is a site [0934 8831] near Saltley Viaduct (Ick, 1842) where a thin bed of peat yielded antlers, thought to be from a species of red deer (*Cervus elephas*). Shotton (1954) reported the remains of horse and red deer within a bed of peat in the alluvium of the River Tame [198 911]; the remains of a Great Irish Deer (*Megaloceras giganteus*), which gave a ^{14}C date of 11 330 ± 140 years BP, were discovered in organic material at the base of the alluvium of the River Tame near Middleton Hall, in the adjacent Lichfield district (Shotton, in Brown, 1980). An organic layer collected from the site of a former gravel pit (now flooded and landscaped), near Lea Marston [211 942], gave a ^{14}C date of 11 700 ± 200 years BP, and yielded a diverse fauna typical of the Zone II (Late Glacial and Post-Glacial, Flandrian age) (Osborne, 1974).

Fauna and flora collected from the River Bourne alluvium, near Shustoke Reservoir, [22 91] have been noted in many publications (Crosskey, 1886; Lapworth, 1913;), culminating in the detailed study by Kelly and Osborne (1964). During excavations for the reservoir in the late 19th century, vertebrate remains were found within a peat (0.7 m thick) which was overlain by red-brown clayey silt (about 2 m thick), and underlain by sand and gravel (about 1 m thick). Kelly and Osborne (1964) sampled two further sites along the banks of the river which had a similar, although attenuated, stratigraphical succession in the alluvium. Extensive studies of the original vertebrate remains, and of the insects and flora, yielded two ages for the peat and organic deposits. The earlier deposits (about 5000 years BP) were laid down in a forested environment dominated by *Alnus* woods on the wet floodplain, and *Tilia-Quercus-Ulmus* woods on the drier hill slopes. The later deposits (about 400 years BP) were laid down when the area was largely devoted to agriculture, and the stream was slower flowing.

Lacustrine alluvium

The Edgbaston Reservoir [04 86] is flanked by a narrow strip of sand and gravel deposited as lacustrine alluvium during periods of high water level. In 1989, during a period of low water level, these deposits were seen to be about 200 m wide.

Peat

Peat has formed within some of the alluvial tracts where impeded drainage has resulted in the rapid accumulation of decaying vegetable matter; some of the occurrences within the alluvium, but not mappable at surface, are noted above (Alluvium).

A small area of peat, up to 0.4 m thick, was proved during site investigation for the M5 motorway, near Newton Road [021 933], but may have been removed during construction of the motorway.

Head

Head, in this district, comprises a mixed deposit of clay- to gravel-grade materials resulting from gravitational solifluction and gelifluction flow under periglacial conditions, but also includes postglacial colluvial deposits. The deposits mapped in the district probably formed in periglacial conditions subsequent to the Anglian, Wolstonian and Devensian cold stages. Head is probably ubiquitous on all of the steeper slopes in the district, but only locally are the deposits of sufficient extent and thickness to be mapped. The deposits vary according to the local source of bedrock or superficial deposit. In the south-west of the district, around Frankley [98 81] and Hunnington [96 82], the Upper Carboniferous bedrock is covered by a thin head deposit (up to 1.5 m thick) consisting of red, and mottled yellow-brown clay with sparse pebbles. The deposit was probably sourced from the topographically higher, drift-free ground (red-brown Salop Formation) located to the south.

In drift-covered areas, locally derived head commonly includes drift lithologies, as beds of gravel, intercalated with pebbly clay, which are difficult to distinguish from in-situ deposits; some of the beds show deformation due to cryoturbation.

On the relatively steep slopes of the Rowley Regis ridge, up to 10 m of head have been recorded at Bury Hill [973 894] (Hutchinson et al., 1973). Here, the deposit consists of grey, yellow, red and variegated silty clay and clayey silt with numerous shear planes (derived from the Etruria Formation bedrock) and boulders of dolerite (derived, up-slope, from the dolerite). The presence of abundant, relict shear planes in the clay matrix renders these deposits inherently unstable and, therefore, susceptible to mass-movement (see Chapter Two).

Horton (1989) recorded the occurrence of an older (Pre-Anglian) head deposit, proved in boreholes, at the base of the Quinton palaeovalley succession (see above).

Landslip

Landslips (see Chapter Two for details) are not common in the district. Small landslips have occurred on the steeper slopes of the Rowley Regis ridge at Bury Hill [972 894] and at Brades Village [977 898] in the west of the district, within head deposits, characterised by shallow shear planes, overlying Etruria Formation mudstone. This landslip is thought to have been initiated as a consequence of overloading due to emplacement of made ground on part of the susceptible slope (Hutchinson et al., 1973). On the northern margins of both Hailstone and Edwin Richards dolerite quarries [88 96], a number of small landslips probably resulted from translational failures on circumferential listric joints within the dolerite body. A larger area of landslip occurs below a steep escarpment in the Etruria Formation at Haden Cross [963 854].

Deep cuttings excavated in the till or glaciofluvial deposits along the course of the Birmingham Canal have failed, in places, as small-scale landslips [008 896, 012 895] involving downslope creep of the drift deposits.

Artificial (man-made) ground

Man has left his impact on the topography of the district in the form of made ground and excavations (worked ground); the latter may be open, backfilled, partly backfilled or flooded. These categories are delineated in more detail on the component 1:10 000 scale maps, and only selected areas comprising the thicker and larger spreads of made ground, and the largest quarries are shown on the 1:50 000 scale map. The potential constraints to development posed by areas of artificial ground are considered in Chapter Two.

MADE GROUND

Made ground comprises areas where the ground material is known to have been deposited by man on top of the natural ground surface, although, in some cases, the top soil and subsoil have been removed first. It includes major embankments, colliery spoil tips, and other significant constructional areas. Only deposits greater than 1.5 m thick are shown on the 1:10 000 scale maps. In addition, made ground is common in built-over urban areas and in this case it has not been delineated unless it forms a topographical feature.

In the historically older parts of the South Staffordshire Coalfield, such as Wednesbury, Darlaston and Oldbury, the original hummocky made ground, predominantly colliery spoil, has been smoothed over during later urban development (Powell et al., 1992. Consequently, the made ground does not form topographical features, and the extent of the deposit is, here, based largely on borehole data. In parts of the old coalfield, where seams such as the Thick Coal were worked at or near surface, the made ground forms a continuous spread, between 2 and 10 m thick, largely composed of colliery shale, sandstone and poor quality coal. More recent deposits of made ground consist of a variety of man-made deposits such as building rubble, pulverised fuel ash, and industrial, chemical and domestic waste. Chemical and domestic waste represent a small proportion of the whole, but are significant since they may produce hazardous toxic leachates or gases (see Chapter Two).

Made ground has been deposited over large areas of the Tame and Rea valleys [for example 09 89 to 14 90], during construction of flood defences, culverting of the water courses and urban development. Other significant areas of made ground shown on the map are the National Exhibition Centre [19 83], Birmingham Airport [16 85], the motor works [15 82] near Elmdon, and site of the former Hams Hall Power Station [21 92].

EXCAVATIONS, OPEN, PARTIALLY BACKFILLED, BACKFILLED OR FLOODED

The categories 'Worked Ground' and 'Backfilled Ground' are shown separately on the component 1:10 000 scale maps, but they are not distinguished on the 1:50 000 scale map, which shows only a single category. Furthermore, only the larger, more significant areas are delineated.

This category of artificial land comprises currently worked and former pits and quarries, and those which have been partially backfilled or backfilled with a variety of waste materials. Also included in this category on the map are areas of workings that have been flooded, such as the aggregate quarries exploiting the alluvium and river terraces along the course of the River Tame from Hams Hall [21 91] to Kingsbury [20 96]. Many of these excavations have been backfilled with pulverised fuel ash derived from the former power station at Hams Hall.

Additional areas of landfill are the former brick pit at Packington [21 85], the sand and gravel workings [22 85, 22 82, 23 81, 22 80] near Meriden and Queslett [06 94], sand pits near Moxley [97 95], former opencast coal workings such as the Patent Shaft site [97 84], dolerite quarries [97 88] near Rowley Regis, and brickclay quarries [98 89] in the Etruria Formation and at [05 78, 20 82] in the Mercia Mudstone Group.

NINE

Structure

The structure of the district has been determined by integrating the results of surface mapping with the interpretation of data from shafts and boreholes, and geophysical surveys. The structural evolution of the district, comprising six major deformation phases, and affecting strata of Cambrian to Jurassic age, is summarised in Table 19.

The district can be divided into three structural provinces (Figure 29). The South Staffordshire Horst, in the west, includes part of South Staffordshire Coalfield, where strata of Ordovician to Triassic age crop out. The Knowle Basin occupies the central part of the district, where strata of Triassic to early Jurassic age crop out. The Coventry Horst in the east, includes the western margin of the Warwickshire Coalfield, where the outcrop consists of strata of late Carboniferous (Westphalian D age). The Knowle Basin represents an extensional Permian to Triassic rift-basin similar in structural style and sedimentary fill to the Worcester Basin (Chadwick and Smith, 1988). The margins of the Knowle Basin are defined by the north-east-trending Birmingham Fault, to the west, and by the north-trending Western Boundary (Meriden) Fault of the Warwickshire Coalfield, to the east.

Four dominant fault orientations, with approximate north, north-west, north-east and west trends, are present in the district (Figure 29). In the South Staffordshire Horst, the Eastern Boundary–Great Barr Fault defines the boundary between the exposed coalfield, to the west, and the concealed coalfield, to the east. The exposed coalfield comprises three areas in which distinct fault orientations are observed. In the north, west-trending faults dominate, generally with throws down to the south, for instance the Darlaston, Lanesfield, Coseley–Wednesbury, Ball's Hill and Tipton and Hill Top faults. In the central part of the exposed coalfield, north-east-trending faults dominate, for instance the Dudley Port Trough faults and Portway Fault. In the south, north-north-west-trending faults dominate, in particular the Russell's Hall Fault. In the concealed coalfield, the faults are dominantly north-trending, with throws down to the east, for instance the Sandwell Park and Ley Hill faults, and the structure of the 'Silurian Bank' (see below and Figure 29). In the Knowle Basin, the main fault orientations, largely determined from geophysical evidence, are north-trending, for instance the Dicken's Heath Fault, with a downthrow to the east, and the Maxstoke Fault, with a downthrow to the west. In the Coventry Horst, the main fault orientations in the Coventry district (Sheet 169) are to the north-west and north (British Geological Survey, 1994).

The complexity of fault orientations in the district results from its position near to the apex of the triangular-shaped Midlands Microcraton (Thorpe et al., 1984; Pharaoh et al., 1987a), and the convergence of three basement domains showing north-east (Caledonian), north (Malvernian) and north-west (Charnian) structural grains. The orientation of many of the major faults affecting the outcrop of rocks of Carboniferous to Triassic age in the district are considered to be inherited from structures present, at depth, in the Midlands Microcraton, which were produced by pre-Carboniferous deformation events (Fraser et al., 1990; Waters et al., 1994).

STRUCTURAL HISTORY

The complex deformation history of the district can be summarised as six distinct tectonic events (Table 19). Evidence for the pre-Westphalian deformation is largely reliant on regional geophysical data, in conjunction with deep borehole and shaft records, since only small outcrops of Lower Palaeozoic rocks are present. Evidence for the late Mesozoic and Cenozoic structural evolution is limited by the absence of rocks of these ages in the district.

Pre-Westphalian deformation

Determination of a pre-Westphalian structural history is hindered by the limited outcrop of rocks of this age. Furthermore, these outcrops occur as fault-bounded lenses along major fault zones which were active during Mesozoic times, such as the Eastern Boundary Fault, in which evidence for pre-Westphalian deformation has been largely obscured by subsequent fault movements. The presence of an unconformity at the base of the Rubery Formation suggests a phase of pre-Llandovery uplift (Smith and Rushton, 1993); this was proved as an erosion surface in the Walsall Borehole (Butler, 1937), in the Lichfield district, and as a slight angular discordance in the Rubery area and in the Redditch district (Old et al., 1991).

Much of the evidence for pre-Westphalian deformation in the district is derived from subsurface information, and in particular gravity and aeromagnetic images (Figure 30, 32), which identify three structural grains within the basement rocks of the Midlands Microcraton (Lee et al., 1990). This Precambrian to Lower Palaeozoic basement complex is bounded to the west by north-east-trending structural lineaments inherited from Caledonian deformation. In the Welsh Borderlands, this dominant structural trend is marked by the Church Stretton and Pontesford–Linley fault zones (Smith and Rushton, 1993). In the Birmingham district, the Birmingham Fault and northern part of the Eastern Boundary Fault may have inherited their orientation from similar

Table 19   Schematic representation of the deformation history within the three main tectonic areas.

KEY LITHOSTRATIGRAPHICAL UNITS AND TECTONIC FEATURES

AGE		SOUTH STAFFORDSHIRE HORST — STRATA	SOUTH STAFFORDSHIRE HORST — EVENTS	KNOWLE BASIN — STRATA	KNOWLE BASIN — EVENTS	COVENTRY HORST — STRATA	COVENTRY HORST — EVENTS
post-Jurassic / Jurassic	Hettangian	no strata on Sheet 168	Major normal faulting; accentuation of horst	no strata on Sheet 168	Major normal faulting; accentuation of graben	no strata on Sheet 168	Major normal faulting; accentuation of horst
Triassic	Rhaetian	Bromsgrove Sst	Angular unconformity, minor inversion of faults e.g. Park Farm Fault	Lias Group; Penarth Group; Mercia Mudstone Group; Bromsgrove Sst; Wildmoor Fm; Kidderminster Fm	Unconformity / angular unconformity?		Similar to South Staffordshire Horst
	Late Anisian and Norian	Wildmoor Sst Fm; Kidderminster Fm					
	Olenekian and Early Anisian	Hopwas Breccia					
	? Olenekian	Clent Fm	Normal faulting e.g. Park Farm Fault	Bridgnorth Sst?	Folding and faulting; uplift and erosion; source of Lower Palaeozoic clasts in Clent Fm at Nechells (Knowle 'Horst' in late Carboniferous to early Permian)		Folding e.g. Hoar Park and Allesley synclines
Permian	? Stephanian &	Salop Formation	Folding and possible reverse faulting e.g. Sandwell Park Fault			Tile Hill Mdst Fm; Meriden Formation	
Carboniferous	Westphalian D	Halesowen Formation	Uplift of Wales–Brabant High, condensed strata in south			Halesowen Formation	
	Westphalian D	Etruria Formation	Erosion, angular unconformity, reverse faulting and/or folding e.g. Russell's Hall Fault		Uplift and formation of structural high; erosion; source of Stockingford Shales in Etruria Fm at Dost Hill	Etruria Formation	Erosion, angular unconformity, reverse faulting and folding, e.g. Arley and Keresley faults and Fillongley Anticline
	Bolsovian	Coal Measures	Synsedimentary faulting, minor normal faulting e.g. Ball's Hill Fault, and reverse faulting e.g. Russell's Hall Fault, Dudley Port Trough			Coal Measures	Minor synsedimentary reverse faulting, e.g. Western Boundary Fault
	Langsettian & Duckmantian						Millstone Grit at Dost Hill
	Namurian; Viséan; Tournaisian		Erosion, marked angular unconformity, regional tilting with dip towards NW				Erosion, marked angular unconformity
Devonian		Downton Group					
Silurian	Pridoli; Ludlow; Wenlock; Llandovery	to Rubery Formation	Erosion, slight angular unconformity			Stockingford Shale Group & Hartshill Sst	
Ordovician	Arenig ?; Tremadoc	Lickey Quartzite & Barnt Green Volcanic Formation		Stockingford Shale Group? & Hartshill Sst?			
Cambrian	Merioneth to Comley						

TECTONIC REGIME

Subsidence — Rifting — E–W extension
Rifting — E–W compression
High subsidence rates — NW–SE to E–W compression
Low subsidence rates — N–S extension — NW–SE to E–W compression
Rifting — N–S extension
Caledonian (Acadian) deformation — NW–SE compression
Thermal subsidence / Tectonic loading
Pulsed Caledonian deformation

MAJOR DEFORMATION PHASES: 6, 5, 4, 3, 2, 1

Vertical ruling denotes non-deposition or erosion of strata

Major unconformities are shown thus: ⌇

Major conglomerate/breccia beds: o o o o o o o o o

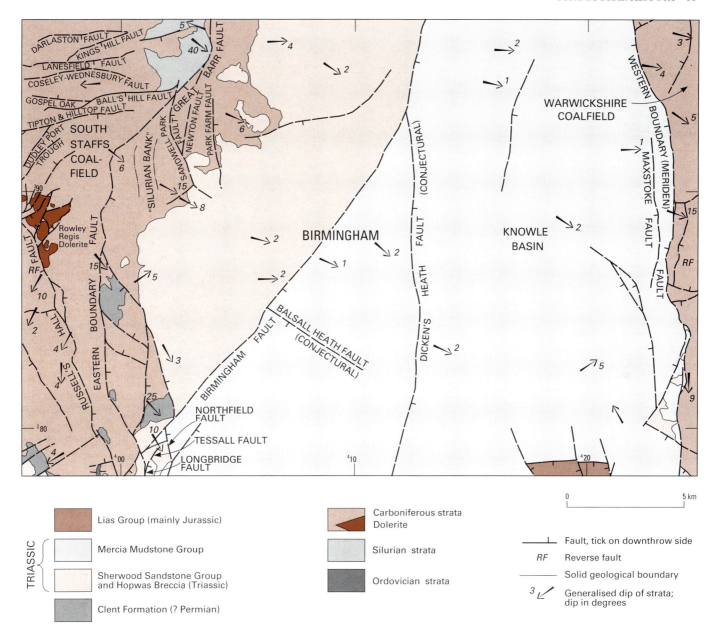

Figure 29 Generalised structural map of the district.

north-east-trending structures present within the micro-craton. Shaft records from the South Staffordshire Horst show that there is a regional tilting of Silurian and Lower Devonian strata, which dip toward the north-west (King, 1921). This tilting is ascribed to the Acadian phase of the Caledonian orogeny (Waters et al., 1994) which occurred during the early to middle Devonian times.

The Midlands Microcraton is bounded, to the east, by north-west-trending structural lineaments. This Charnian grain is parallel to fold axes of Proterozoic age which are present in exposures of Charnian rocks along the edge of the microcraton. The Charnian grain is considered to have been reactivated during Lower Palaeozoic times.

North-west-trending 'eastern Caledonide' structures to the east of the microcraton are present within thick,

basinal Cambrian and Silurian sedimentary rocks, metamorphosed to greenschist facies (Pharaoh et al., 1987b; Lee et al., 1990). In the district, the Russell's Hall Fault may have inherited its orientation from a Charnian structure present within the Midland Microcraton.

North-trending faults, such as the southern part of the Eastern Boundary Fault of the South Staffordshire Coalfield, the Dicken's Heath Fault and the Western Boundary Fault of the Warwickshire Coalfield, parallel a Malvernian basement trend in the microcraton associated with the Malvern line. This fault zone has been shown by petrological, geochemical and structural studies to represent the boundary between two Precambrian island arc terranes (Pharaoh et al., 1987b).

Carboniferous to Early Permian deformation

LANGSETTIAN TO BOLSOVIAN DEFORMATION

During the late Carboniferous, the district was located at the southern margin of the Pennine Basin, which was bounded to the south by the Wales–Brabant High (see Chapter Five). Although several tectonic models have been proposed to explain the origin of the Pennine Basin, the model of Leeder (1982) is broadly accepted, with modifications discussed in detail by Waters et al. (1994). Leeder (1982) proposed that, in central Britain, a phase of early Carboniferous, north–south extension occurred in response to back-arc stretching adjacent to northward-directed Bretonic/Ligurian subduction. Lower Carboniferous rocks are not present in the district, probably because the area was land, and consequently there is no evidence of this extensional deformation phase. Subsequent to plate collision and cessation of subduction, the rift basins evolved into a single, thermally subsiding, Pennine Basin (Leeder, 1982). In central Britain this is manifested in a broad 'bull's-eye' isopachyte pattern for post-extension Namurian and Westphalian strata (Trueman, 1947; Fraser et al., 1990), and a southward thinning and onlap of these strata toward the Wales–Brabant High (Fulton and Williams, 1988; Besly, 1988).

Although thermal subsidence probably acted as the major control on sedimentation of the Coal Measures and Etruria Formation during Langsettian to Bolsovian times, evidence from the South Staffordshire Coalfield (Waters et al., 1994) and Warwickshire Coalfield (Fulton and Williams, 1988) shows that tectonic controls also had a local influence on sedimentation. In the district, synsedimentary normal and reverse faulting has been proved for a number of structures (e.g. Russell's Hall Fault) in the South Staffordshire Horst. Pulses of broadly north-south extension resulted in the activation of west-trending structures as normal faults, whereas pulses of east–west compression activated north–west- to north–east-trending structures as reverse faults (Waters et al., 1994).

In the north of the South Staffordshire Horst, west-trending faults dominate, generally with throws down to the south (Figure 29). The main faults, with estimates of maximum throw, include the Darlaston (20 m), Lanesfield (55 m), Coseley–Wednesbury (65 m), Ball's Hill (130 m) and Tipton and Hill Top (55 m) faults. Synsedimentary activity on west-trending normal faults is indicated by the interseam thickness between the lower leaf of the Thick Coal and the Flying Reed Coal. Isopachyte trends define a broadly west-trending, asymmetric zone of locally thick strata parallel with the small half-grabens bound by the Tipton–Hill Top, Coseley–Wednesbury and Ball's Hill faults (Waters et al., 1994, fig. 8; Plate 4). The Gospel Oak Fault, a west-trending fault, which downthrows to the north, appears to have controlled the onset of synsedimentary red-bed formation, which marks the change from Coal Measures to the Etruria Formation. South of the fault, in the footwall, the red-beds commence about 2 m above the Two Foot Coal, whereas in the hanging wall, reddening

occurs about 20 to 25 m above this coal. This suggests that, during sedimentation, the footwall formed a topographic high (well-drained conditions) relative to the hanging wall (Waters et al., 1994).

A variation in the thickness of Coal Measures (Langsettian to Bolsovian) across the Russell's Hall Fault, in the exposed coalfield, is also interpreted as reflecting synsedimentary faulting. Thinner, condensed, Coal Measures successions, with coal seams either thin or absent, are present to the east (hanging wall) of the Russell's Hall Fault which is interpreted as having been an active reverse fault at this time (Waters et al., 1994). 'Silurian Bank' (Eastwood et al., 1925) is a term used to denote north-trending basement highs of Silurian strata in the Sandwell–Handsworth area (Figures 29, 30); Coal Measures strata onlap and coal seams thin against these highs which may have been bounded by active faults.

A north-north-east-trending zone of locally thickened Coal Measures strata between the New Mine and Thick coals occurs parallel with, but to the south-east of, the graben of the Dudley Port Trough (Figure 29). The presence of the thick sequence in the footwall south-east of the graben may indicate the presence of synsedimentary reverse displacement on this fault (Waters et al., 1994). This is supported by the observation that, regionally, the onset of primary reddening generally occurred a few metres above the Two Foot Coal, with the exception of this north-north-east-trending zone, located to the south-east of the Dudley Port Trough. Here, a topographic low, associated with high water table levels, persisted in the footwall of the south-east fault of the Dudley Port Trough (Waters et al., 1994). The present geometry of this structure is a graben, with a maximum throw on the eastern fault of 131 m, down to the west-north-west, is probably the result of renewed normal movement during post-Carboniferous extension.

On the Coventry Horst, evidence for synsedimentary faulting occurs largely in the adjacent Coventry (Sheet 169) district and is discussed fully by Bridge et al., 1998. Syndepositional movement on the north-north-east-trending Arley Fault is interpreted as resulting in local facies variations in both the Millstone Grit and Coal Measures (Taylor and Rushton, 1971). A seam split in the Nine Feet seam of the Thick Coal, which occurs parallel with, and to the east of, the north-trending Western Boundary Fault (Bridge et al., 1998), suggests that this fault was active as a reverse fault during Duckmantian times. Furthermore, the presence of a major sandstone washout channel, at the level of the Two Yard seam, oriented north-north-west, subparallel with the trend of the Western Boundary Fault, suggests a degree of structural control on the orientation of the major fluvial channels. The north-north-west trend of isopachytes for Langsettian and Duckmantian strata is interpreted as reflecting a depositional control imposed by the reactivation of Charnian basement faults with a similar trend (Fulton and Williams, 1988). Some of these structures have been interpreted as showing synsedimentary reverse displacements (Corfield, 1991).

BOLSOVIAN TO WESTPHALIAN D DEFORMATION

This was marked by a phase of fault inversion and regional uplift of both the present-day South Staffordshire and Coventry horsts, resulting in the development of an angular unconformity beneath the Halesowen Formation. Uplift (reverse movement) of the present-day Knowle Basin along major bounding faults, such as the Western Boundary Fault and the Birmingham Fault resulted in a structural high in this area (see below). Deformation resulted from east–west to north-west–south-east compression, with most of the active faults and folds having a north-north-west–north-north-east trend. This deformation is interpreted as the first phase of Variscan (Hercynian) deformation which continued to affect the region until Early Permian times.

In the South Staffordshire Horst, rapid lateral thickness variations in the Etruria Formation, present to the west of the Russell's Hall Fault, indicate faulting and/or folding and erosion prior to deposition of the Halesowen Formation. The Russell's Hall Fault was active as a reverse fault at this time; a prominent drag fold present to the west, in the footwall, of the fault indicates syndepositional deformation. Locally, the Russell's Hall Fault comprises a fault zone up to 800 m wide, with a throw in the Etruria Formation of 46 m, down to the south-west. Further folds generated during this phase of compression are present in the adjacent Dudley (167) district, including the north-north-east-trending Netherton Anticline and north-trending periclines of Castle Hill, Wren's Nest and Hurst Hill (Waters et al., 1994).

In the Knowle Basin, the evidence for this phase of deformation is limited to the Trickley Lodge Borehole [SP 1603 9884], located about 2 km north of the district. Here, the Halesowen Formation rests unconformably upon Cambrian basement rocks, indicating that, prior to deposition of the Halesowen Formation, the Knowle Basin formed an uplifted block, subjected to erosion. Earlier uplift in Etruria Formation time is supported by the presence in the Etruria Formation at Cliff Quarry [SP 220 988], Dosthill, and other parts of the Warwickshire coalfield (Barrow, 1919), of abundant subangular 'shale-chip' clasts derived locally from the Cambrian Stockingford Shale Group.

In the Warwickshire Coalfield (adjacent Coventry district, Sheet 169), folding of the north-east-trending Fillongley Anticline, and displacements on the north-east-trending Arley Fault and north-north-west-trending Keresley reverse fault, also occurred prior to deposition of the Halesowen Formation (Old et al., 1990; Bridge et al., 1998).

WESTPHALIAN D TO EARLY PERMIAN DEFORMATION

Following the phase of deformation associated with the development of the sub-Halesowen unconformity, a period of subsidence was re-established in the northern part of the district. In the north of the South Staffordshire Horst, thickness data for the Salop and Clent formations indicate an increase in subsidence rates compared to Langsettian to Bolsovian times; this was possibly a response to the establishment of the region

within the foreland basin of the Variscan fold-thrust belt, which was located to the south of the Wales–Brabant High (Besly, 1988). In contrast, the sequence recorded in Daleswood Farm Borehole [SO 9511 7913] (Glover and Powell, 1996) suggests that the Salop Formation is condensed in the south of the Pennine Basin, reflecting local, tectonically induced uplift of parts of the Wales–Brabant High. This culminated in the development of an unconformity at the base of the Clent Formation. Uplift of the South Staffordshire Horst was controlled, to the east, by the Eastern Boundary Fault.

Evidence of tectonic activity on individual structures at this time is mostly inferred from provenance data. The presence of Carboniferous limestone pebbles in the Salop Formation (Chapter Five), Lower Palaeozoic (including Cambrian), and Neoproterozoic Uriconian and 'Charnian' pebbles in the Clent Formation in the south of the district (King, 1893, 1921) and in boreholes (e.g. Nechells Gasworks No. 1 Borehole) in central Birmingham (Boulton, 1924; 1928; 1933), suggests local provenance from, fault-bounded highs situated adjacent to the late Carboniferous to Early Permian basins (see Chapter Five). Direct evidence of syndepositional movement has been largely obscured by subsequent, large normal displacements on the major faults during Mesozoic times. A rare example of deformation, possibly of this age, occurs in the concealed coalfield, to the east of the Eastern Boundary Fault. Here, north-east-trending to north-north-east-trending anticlines and synclines, with dips on the limbs of up to 12° are proved, at depth, in the Coal Measures in mine workings from Hamstead Colliery. A series of tight folds, with a vergence to the west is also present in the Enville Member (Salop Formation), in the hanging wall of the north-trending Sandwell Park Fault [023 926]. Tight folds are not seen in the overlying Triassic strata, and it seems likely that deformation occurred during a phase of reverse displacement on the dominantly north-trending faults, such as the Eastern Boundary and Sandwell Park faults, during late Carboniferous to Permian times.

In the Coventry Horst, this phase of deformation produced the Hoar Park and Allesley synclines which dominate the structure of the Warwickshire Coalfield in the Coventry district (Bridge et al., 1998).

Permian to Triassic deformation

During Permian times, there was a change in the style of deformation from pulses of broadly east-west, Variscan compression or transpression to a phase of widespread rift-controlled subsidence, associated with east–west extension. A series of graben and half graben, including the Worcester, Stafford and Cheshire basins, developed in response to the effects of extension on north-trending (Malvernian) basement structures (Chadwick, 1985; Glennie, 1990). This phase of extension is evident in the adjacent Dudley district (Sheet 167) west of the South Staffordshire Horst, and resulted in the formation of the Bratch Trough and deposition of aeolian sand (Bridgnorth Sandstone). This is a north-trending graben representing a northern continuation of the Worcester Graben. Subsidence of the Knowle Basin also commenced at this time (see below). The extent of Permo-Triassic displacement on faults is difficult to distinguish from that associated with post-Triassic extension, because of the absence of Jurassic strata over most of the district.

In the northern part of the Knowle Basin (Trickley Lodge Borehole), the Triassic Sherwood Sandstone Group rests unconformably upon Carboniferous strata which had been faulted and folded during the end-Carboniferous Variscan phase of deformation. Farther south, within the district, the stratigraphy and structure of the Knowle Basin is poorly known. The east–west cross-section shown on Birmingham Sheet 168 is modelled from knowledge of the structural style, and basin-fill in the Worcester Basin (Chadwick and Smith, 1988), evidence of thick basin-fill (Sherwood Sandstone Group) proved in the Blythe Bridge Borehole [2119 8979] (Figure 21), the succession proved in the Trickley Lodge Borehole [SP 1603 9884] where the Halesowen Formation Formation overlies Lower Palaeozoic rocks, and from provenance studies of pre-Triassic (Salop and Clent) formations and the early Triassic Hopwas Breccia. The cross-section shows a thick Permian (equivalent to the aeolian Bridgnorth Sandstone) and Triassic succession overlying Lower Palaeozoic or Neoproterozoic strata which were formerly uplifted during late Carboniferous to Early Permian times. Triassic rifting and subsidence, during an extensional regime, resulted in the formation of the Knowle Basin and the deposition of a thick sequence of predominantly alluvial sediments in Early Triassic (Scythian) times.

An intra-Triassic unconformity (equivalent to the Hardegsen Unconformity; Geiger and Hopping, 1968; Wills, 1970a; Evans et al., 1993) is present locally at the base of the Bromsgrove Sandstone. This unconformity is well developed in the vicinity of Harborne, between the Eastern Boundary and Birmingham faults, where the Bromsgrove Sandstone oversteps the Wildmoor and Kidderminster formations to rest upon the Clent Formation (Eastwood et al., 1925; Wills, 1970). The presence of this unconformity in the Knowle Basin had also been postulated by Wills (1976). In the eastern part of the Knowle Basin, the Dumble Farm Borehole (Figure 21) proves the Bromsgrove Sandstone resting unconformably upon the Meriden Formation, the Kidderminster Formation and Wildmoor Sandstone being absent. The north-trending Park Farm Fault, at Hamstead, shows a normal throw in Coal Measures strata of about 106 m, down to the west, but a reverse throw at the base of the Triassic Kidderminster Formation, of 5 to 10 m, down to the east. This suggests a major phase of extension prior to Triassic times (possibly Permian) and minor compression and fault inversion, during or after Triassic times, possibly associated with the intra-Triassic unconformity.

The unconformity at the base of the Bromsgrove Sandstone may have developed in response a reduced rate of subsidence in the Knowle Basin, associated with isostatic re-equilibration that took place between the main Early Triassic (Scythian) phase of lithospheric extension (rifting), and the subsequent phase of thermal relaxation (McKenzie, 1978; Chadwick, 1985; Evans et al., 1993).

Post-Triassic deformation

The final phase of deformation comprises a major phase of east–west extension, resulting in the accentuation of the major horst and graben structures of central England (Hains and Horton, 1969), and, in particular, the north-trending South Staffordshire and Coventry horsts and the Knowle Basin, of the district. A number of reverse faults of Variscan age may have been reactivated as normal faults at this time. For example, the Russell's Hall and Sandwell Park faults currently show a net normal displacement, but had been reverse faults during Variscan deformation (see above). It is generally not possible to distinguish between the amount of displacement in post-Triassic times from displacement that occurred in Permo-Triassic times.

This phase of deformation occurred after the deposition of the Triassic Mercia Mudstone Group, and is believed to be post-Triassic in age, although the precise timing cannot be ascertained because of an absence of Jurassic or younger rocks over most of the district.

The western bounding fault of the Knowle Basin, the **Birmingham Fault**, juxtaposes the Mercia Mudstone Group, to the east, against the Sherwood Sandstone Group, to the west. The throw on the Birmingham Fault, down to the east, is estimated to be about 200 m in the Erdlington–Gravelly Hill area [10 90 to 11 90], diminishing rapidly to about 60 m, to the south-west of the junction with the north-trending Northfield Fault [019 800]. The Birmingham Fault was formerly exposed at Tessall Lane [006 785], where it comprises a complex zone of disturbance in the Alveley Member (Salop Formation), including at least four fault planes (Wills and Shotton, 1938).

The **Longbridge Fault**, in the west of the Knowle Basin has a downthrow to the east of about 200 m, juxtaposing Mercia Mudstone Group against Kidderminster Formation. The fault was formerly exposed at Tessall Lane [009 784], forming a 200 m wide belt of seven faults, including the **Tessall Fault** (Wills and Shotton, 1938). The presence of two slices of Salop Formation caught up in the fault zone suggest a complex history of faulting. The throw on the Longbridge Fault decreases toward the north, and at Frankley Lodge Farm [006 797] it splits into three faults, the eastern branch being the Eastern Boundary Fault. The **Eastern Boundary Fault** juxtaposes the Mercia Mudstone Group against Carboniferous strata in the adjacent Lichfield (154) district. The throw, down to the east, increases northward from about 80 m, near Oldbury [000 918], to about 550 m, in the vicinity of Hamstead [034 950] where it is named the **Great Barr Fault**. This northward increase in throw represents the accumulation of throws, also down to the east, on a series of north-trending splay faults, including the Sandwell Park Fault. To the east of the Eastern Boundary Fault, the Sherwood Sandstone Group and underlying strata have been folded into a broad, north-trending, asymmetric syncline, with a steeper western limb forming the scarp of Barr Beacon and with dips of up to 12° recorded at Pinfold Lane Quarry [059 967]. The syncline is interpreted as developing in the hanging wall of the Eastern Boundary Fault in response to post-Triassic displacements on the fault.

The eastern bounding fault of the Knowle Basin, the **Western Boundary (Meriden) Fault** of the Warwickshire Coalfield, juxtaposes Mercia Mudstone Group, to the west, against Carboniferous strata to the east. The throw on this fault is estimated as 550 m, down to the west.

Within the Knowle Basin few faults have been detected by surface mapping, because of the general absence of suitable lithological markers in the Mercia Mudstone Group. The main faults of this age which have been identified, in part from geophysical evidence, include the north-trending Dicken's Heath and Maxstoke faults. The **Dicken's Heath Fault**, proved between the Bromford Lane Borehole [1176 8919] and Birmingham Race Course Borehole [1282 8970], downthrows the base of the Mercia Mudstone Group about 90 m to the east (Piper, 1993b); the throw on the fault decreases northwards. The **Maxstoke Fault** is estimated to have a throw, down to the west of about 330 m, within Triassic strata. Within the Knowle Basin, the Mercia Mudstone Group, at outcrop dips about 1 to 2° to the east or south-east; small faults have been mapped, in drift-free terrain, on the basis of offset on features formed by skerries or the Arden Sandstone Formation.

GEOPHYSICAL SUMMARY AND REGIONAL GEOPHYSICAL TRENDS

The regional gravity and magnetic data provide valuable structural information about the district and the surrounding area. The wide variety of rocks present and their greatly varying physical properties give rise to many large Bouguer gravity and magnetic anomalies. Modern processing and imaging techniques also allow more subtle anomalies to be studied. Many of the Bouguer gravity anomalies seen in and around the district are attributable to thickness variations in the Triassic and Carboniferous strata, while others reflect structures within the Lower Palaeozoic and Precambrian. Further details of geophysical studies in and around the district can be found in Royles (1996).

Physical properties

Physical property determinations have been made in the English Midlands and Welsh Borderlands by several workers. During this study these data were augmented by evaluations of geophysical logs from boreholes. Values considered to be appropriate for the Birmingham district are detailed in Table 20.

Sources of data

The regional geophysical data for the district and surrounding area comprise gravity and aeromagnetic data from the national BGS databanks. The regional

gravity data coverage was generally good, except for an area of 35 km^2 and this area was subsequently improved during the present study by a survey carried out by the Department of Earth Sciences of Birmingham University, under the supervision of W H Owens. The aeromagnetic data were acquired in 1955 in two tranches. The area north of Northing ^{2}98 was flown at a mean terrain clearance of 305 m, and the area south of this line was flown at a constant barometric height of 549 m. The flight-lines were east–west, 1 mile (1.6 km) apart with north–south tie-lines 6 miles (9.7 km) apart.

Data presentation and interpretation

The principal features of the gravity and magnetic data are illustrated by the maps in Figures 30, 31 and 32 with Bouguer gravity and magnetic anomalies labelled G and M respectively and the gravity lineaments labelled GL (Figure 31). The Bouguer gravity and magnetic anomalies along Generalized Horizontal Section 1 on the 1:50 000 Series Birmingham Sheet 168 are shown in Figure 33 and have been modelled using GRAVMAG (Busby, 1987; Pedley, 1991) to illustrate the relationship between the interpreted geology and the potential field anomalies and to suggest possible geometric forms for the deeper bodies within the Neoproterozoic (Precambrian) basement.

The Bouguer gravity anomaly field contoured at intervals of 1 milligal is shown in Figure 30; while in Figure 31, the first horizontal derivative of the same field (i.e. gravity gradient) is displayed as a pseudo-relief illuminated by an imaginary light source located in the west, emphasising short wavelength anomalies due to near-surface density contrasts. It is particularly effective in highlighting contiguous zones of density contrast such as faults. Further details of these techniques are given in Lee et al. (1990) who also describes a large part of central Britain, including this district. Several pseudo-relief images have been used in this interpretation in order to select structural lineaments.

The aeromagnetic data used to produce Figure 32 were processed in order to remove the asymmetry due to the inclination of the geomagnetic field. The anomalies produced by these reduced-to-pole data should directly overlie the source bodies.

The three structural provinces of the district, the South Staffordshire Horst, the Knowle Basin and the Coventry Horst are well defined geophysically, particularly by the Bouguer gravity anomalies (Figure 30).

South Staffordshire Horst

The South Staffordshire Horst is represented by a major Bouguer anomaly high (G1), with a maximum value of 17 milligals just to the north of the district, and a major magnetic anomaly (M1) with its maximum (500 nanotesla) lying on the district margin. The near-coincidence of these anomaly highs can also be seen along the magnetic and Bouguer gravity anomaly profiles (Figure 33), and the similarity of their general form suggests that they could be derived from a dense magnetic body, but are perhaps more likely to be due to separate but

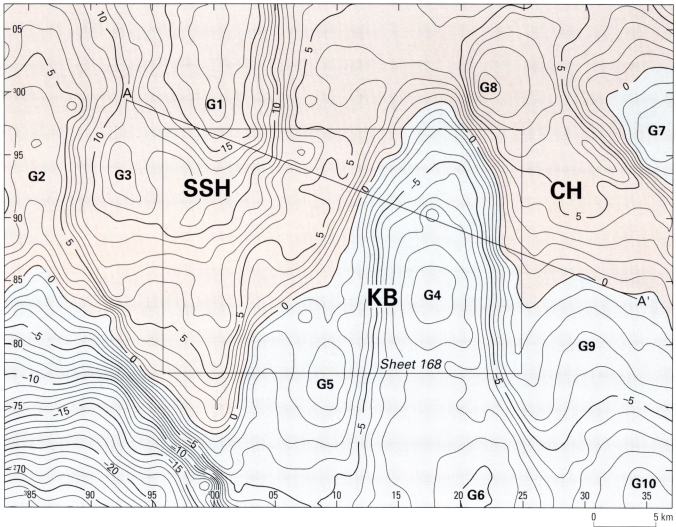

Figure 30 Bouguer gravity anomaly map of the district and adjoining areas. A variable Bouguer reduction density has been used. Anomalies G1–10 are discussed in text. Line AA′ is line of anomaly profile shown in Figure 33.

SSH South Staffordshire Horst, KB Knowle Basin, CH Coventry Horst.

intimately associated bodies. In Figure 33 these coincident highs are modelled by a single polygon with high density and magnetic susceptibility. It fails to produce a perfect fit, but does illustrate the close relationship between the Bouguer gravity and magnetic highs, and also that it is here that the source of the magnetic anomalies beneath the South Staffordshire Horst is closest to the surface.

The boundaries of the South Staffordshire Horst are well defined by steep gravity gradients (Figure 31) and, to a lesser extent, by the magnetic gradients (Figure 32). GL1, located to the west of the district, is related to the Western Boundary Fault of the Staffordshire Coalfield. This lineament can be traced south as GL2a and GL2b, which mark the eastern margin of the Worcester Basin, GL2a corresponding to the Lickey End Fault. At its northern end GL1 marks the position of the Lloyd's House and Stapenhill faults rather than the Western Boundary Fault. To the west of the Stapenhill Fault,

the Bratch Trough, a northward continuation of the Worcester Basin, can be seen as a minor Bouguer anomaly low (G2, Figure 30). The Eastern Boundary Fault of the South Staffordshire Coalfield is marked by GL3 (Figure 31), but it is GL4 which is the more prominent gradient, marking the eastern margin of the concealed coalfield. The northward continuation of the Eastern Boundary Fault, the Great Bar Fault, is not generally associated with a steep gravity gradient, even in the region where Silurian rocks juxtapose Carboniferous strata. This can also be seen on the Bouguer anomaly profile in Figure 33 where the modelled steep gradient (at 11 km) at the contact fails to fit the observed curve. This may be due in part to rather sparse data coverage in the area, but could indicate a smaller density contrast between the Silurian and Carboniferous rocks than would be expected, or that the anomaly is being masked by deeper structures. At its southern extreme, where it meets the Lickey End Fault (GL2a, Figure 31) (in the

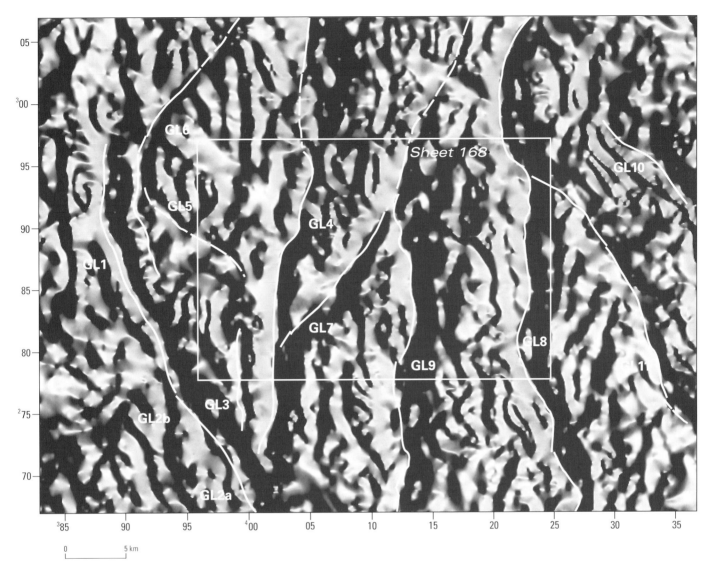

0   5 km

Figure 31 Shaded pseudo-relief map of the first horizontal derivative of the Bouguer gravity
field (i.e. gravity gradients) illuminated from the west.
Bold white lines (labelled GL 1 to 11) are lineaments discussed in text.

Redditch district), GL4 marks the position of the Longbridge Fault. The Russell's Hall Fault, GL5 (less
prominent than some other major lineaments due to the
western illumination) and GL6, the Western Boundary
Fault, form an arcuate feature in Figure 31 which correlates generally with the zone of maximum gradients of
the magnetic anomaly (Figure 32) in the north-western
corner of the district. A small positive magnetic anomaly
(M2, Figure 32) [SS 9400 9200], just west of the sheet
boundary, coincides with a Bouguer gravity anomaly high
(G3, Figure 30) over the Dudley Ridge where the Lower,
Middle and Upper Elton formations crop out.

The South Staffordshire Horst, as defined earlier in
this chapter, is bounded to the south-east by the north-
east-trending Birmingham Fault which is marked by a
prominent lineament GL7 (Figure 31). In the area
between GL4 and GL7 (part of the concealed coalfield)

the Bouguer gravity anomaly field is some 10 milligals
lower than that seen in the area 10 km to the north-west.
This is well illustrated by the profile in Figure 33, where a
distinct change in gradient can be seen at 14 km on the
Bouguer gravity anomaly profile, the gradient between
14 and 19 km being relatively low. Coincident with this
zone of lower Bouguer gravity gradient is a large lenticular magnetic anomaly (M3, Figure 32), orientated south-
west–north-east and constrained to the south-east by the
Birmingham Fault. In Figure 33 the Birmingham Fault
can be seen to be closely associated with the interpreted
deep magnetic discontinuity. The association of this
magnetic anomaly with lower Bouguer anomaly values
suggests that the source could be somewhat different
from that causing M1 (Figure 32) and G1 (Figure 30) or
that in this region there exists a lower density, non-
magnetic body, as modelled by B3 in Figure 33. To the

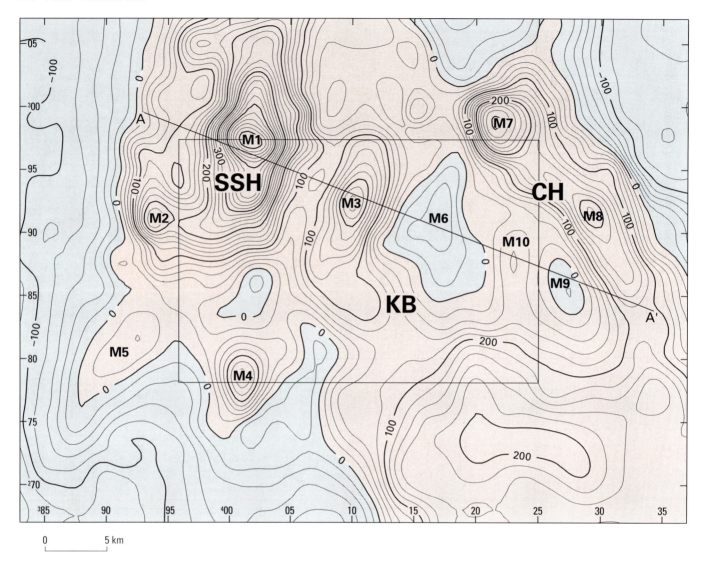

Figure 32 Magnetic anomaly map of the Birmingham district and adjoining areas. Anomalies M1–10 are discussed in text. Line AA′ is line of anomaly profile shown in Figure 33. The anomalies have been reduced-to-pole.

SSH South Staffordshire Horst, KB Knowle Basin, CH Coventry Horst.

south-west, just west of the area in which the Birmingham Fault meets the Eastern Boundary Fault, there is a small asymmetric magnetic anomaly (M4, Figure 32) with steep gradients aligning with the Eastern Boundary Fault (GL3, Figure 31) and the Birmingham Fault (GL7, Figure 32). The two anomalies (M3 and M4, Figure 32) can be seen on shaded pseudo-relief images of the magnetic anomalies to form a continuous magnetic high lying parallel to the Birmingham Fault (GL7, Figure 31), thus supporting the theory that the fault and the source of the anomalies are related. Just west of the district a ridge of high magnetic anomaly (M5, Figure 32) is again oriented south-west–north-east and is associated with a subtle change in the Bouguer gravity anomaly gradients. The broad nature of the feature suggests that its source lies at depth in the sub-Mesozoic basement beneath the northern extremity of the Worcester Basin.

Knowle Basin

The Knowle Basin is delineated by the Birmingham Fault (GL7, Figure 31) to the north-west and by the Western Boundary Fault of the Warwickshire Coalfield (GL8, Figure 31) and the Maxstoke Fault to the east. The Knowle Basin can be seen in Figures 30 and 31 to be dissected by a major north–south lineament (GL9) which corresponds

Figure 33 Magnetic anomaly profile, Bouguer anomaly profile and theoretical model along line AA′. (Magnetic profile is shown with a regional field of 80 nanotesla added.) See Figure 32 for location of profile.

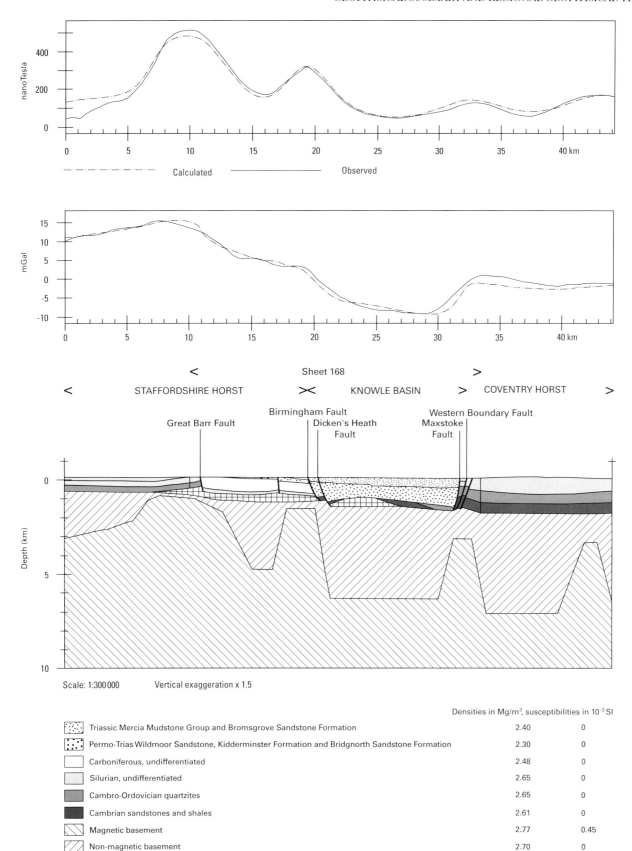

nanoTesla

400

200

0

0   5   10   15   20   25   30   35   40 km

— · — · — Calculated   ————— Observed

mGal

15
10
5
0
-5
-10

0   5   10   15   20   25   30   35   40 km

< Sheet 168 >

< STAFFORDSHIRE HORST >< KNOWLE BASIN > COVENTRY HORST >

Great Barr Fault

Birmingham Fault
Dicken's Heath
Fault

Western Boundary Fault
Maxstoke
Fault

Depth (km)

0

5

10

Scale: 1:300 000   Vertical exaggeration x 1.5

Densities in Mg/m³, susceptibilities in 10⁻³ SI

		Density	Susceptibility
	Triassic Mercia Mudstone Group and Bromsgrove Sandstone Formation	2.40	0
	Permo-Trias Wildmoor Sandstone, Kidderminster Formation and Bridgnorth Sandstone Formation	2.30	0
	Carboniferous, undifferentiated	2.48	0
	Silurian, undifferentiated	2.65	0
	Cambro-Ordovician quartzites	2.65	0
	Cambrian sandstones and shales	2.61	0
	Magnetic basement	2.77	0.45
	Non-magnetic basement	2.70	0
	Low density non-magnetic basement	2.60	0

with the Dicken's Heath Fault; to the east is a major Bouguer gravity anomaly low of -10 milligals (G4, Figure 30). Very little is known about the nature of the strata beneath this area due to the absence of deep boreholes and seismic data. The Bouguer gravity low was interpreted by Cook et al. (1955) to be due to approximately 300 m of Triassic sedimentary rocks overlying some 1500 m of Coal Measures. This has since been re-evaluated and the preferred geological model for the Knowle Basin is of up to 1500 m of Permo-Triassic strata resting upon Cambrian and Precambrian rocks. Unfortunately, the similarity of densities and uncertainty of thicknesses for Coal Measures and Permo-Triassic rocks prevent the use of gravity modelling in confirming the absence of Coal Measures. The Bouguer gravity anomaly coincides, in part, with a magnetic anomaly low (M6, Figure 32), leading Busby et al. (1993) to attribute part of the anomaly to a lower density body within the Precambrian.

The area bounded to the east by the Dicken's Heath Fault (GL9, Figure 31), to the north-west by the Birmingham Fault (GL7) and to the west by the Longbridge Fault (GL4) is a Bouguer anomaly shelf characterised by slack gradients and anomaly values intermediate between those of the Staffordshire Horst and the Knowle Basin low. At the southern end of this shelf, in the Redditch district, at least 600 m of Upper Carboniferous and Lower Permian sedimentary rocks overlie Lower Palaeozoic strata (Old et al., 1991), possibly thinning eastwards onto a basement ridge, producing the small positive anomaly (G5, Figure 30) which is coincident with a slight rise in magnetic anomaly values.

The Dicken's Heath Fault continues south into the Redditch district, while the Western Boundary Fault of the Warwickshire Coalfield (GL8, Figure 31) swings slightly east of south, following the trend of the Warwick and Whitnash faults in the Warwick district. The southward continuation of the Knowle low, with a comparable Bouguer anomaly minimum (G6, Figure 30) suggests that an equally thick Permo-Triassic succession may be present here, as farther north, but this is by no means unequivocal as another low (G10), 10 km east of G6, within the Coventry Horst structural province (in the Warwick district), is due to a thick Carboniferous sequence (Old et al., 1987).

Coventry Horst

The Coventry Horst (Figure 30) is delineated in the west by the Western Boundary Fault of the Warwickshire Coalfield, and the Maxstoke Fault (GL8, Figure 31). To the north-east, outside the district, a series of parallel lineaments (Figure 31) reflect outcropping Cambrian and Neoproterozoic (Precambrian) rocks of the Nuneaton inlier. North-east of these the Poyntington Fault (GL10) marks the boundary between the Coventry Horst and the Hinckley Basin where over 500 m of Permo-Triassic strata are believed to be present (Worssam and Old, 1988; Bridge et al., 1998). At its northern end the Coventry Horst terminates in a major Bouguer anomaly high (G8, Figure 30) which coincides with magnetic anomaly M7 (Figure 32), and which continues to the south-east for

about 15 km as a ridge of high values (M8). This ridge is particularly notable for the way in which its trend reflects that of the north-eastern and eastern margin of the Coventry Horst. The ridge trends parallel with the associated Bouguer gravity anomaly (G7), but is offset by about 5 km to the south-west. The south-western margin of this magnetic ridge is marked by a gravity lineament (GL11, Figure 31). This magnetic ridge can be seen in Figure 33 but the profile crosses the ridge at a comparatively low point (see Figure 32) so that the interpreted causal structure would be much larger to the north. In the region of higher Bouguer anomaly values, close to the Western Boundary Fault, is a small positive magnetic ridge (M10, Figure 32) interpreted as another deep magnetic structure likely to have played a key role in the formation of the Knowle Basin. The modelled Bouguer gravity anomaly profile in this area (Figure 33) does not give a precise fit but this may be due to the profile crossing the Western Boundary Fault at a rather narrow angle. Between GL11 (Figure 31) and the Western Boundary Fault (GL8) the Bouguer anomaly contours form a small trough (G9, Figure 30) reflecting the thickening of Westphalian rocks southwards, and coincident with a magnetic anomaly low (M9, Figure 32).

The close correlation of magnetic anomalies with geological structures throughout the Coventry Horst suggests that the magnetic body or bodies responsible have played an important part in the structural history of the province.

Sources of magnetic anomalies

The district lies at the northern extremity of a major magnetic high, some 30 km wide and extending 100 km to the south, which was interpreted by Wills (1973, 1978) as indicating deeply buried Precambrian rocks. Evidence from the Withycombe Farm Borehole (Cornwell, in Poole 1978; Rushton and Molyneux, 1990) and studies by Pharaoh et al. (1987 a, b) led Lee et al. (1991) to suggest that the magnetic ridge might represent the plutonic core of a Charnian volcanic arc embedded within the Midlands Microcraton. In the Welsh Borderlands, magnetic anomalies have been shown to be closely associated with the distribution of near-surface Uriconian (Neoproterozoic) rocks (Brooks and Fenning, 1968). Some 10 km to the north-west of the district (in the Wolverhampton district) the Heath Farm Borehole [SJ 9330 0940] proved Uriconian basement at a depth of 1565 m below OD. Magnetic susceptibility measurements made during this study gave variable, but locally high readings (>10^{-2} SI) indicating that the basement in the Birmingham district could also be Uriconian rocks, and be the source of the magnetic anomalies observed.

The minimum depth to the source of the magnetic anomaly high in the Coventry Horst occurs close to a small inlier at Dosthill, comprising the Stockingford Shale Group intruded by lamprophyre dykes and sills (Taylor and Rushton, 1971). Geophysical data from the Coventry district (Bridge et al., 1998) indicate that parts of these differentiated lamprophyre sills have high magnetic susceptibilities. Likewise, measurements of hornblende-rich components of thicker differentiated sills within the Stockingford Shales at Nuneaton produce equivalent values. Rocks with this amount of magnetite could give rise to the observed anomalies if they occurred in sufficient volume.

The Nuneaton and Dosthill intrusions, along with similar rocks present in the Malvern Hills and near the Wrekin in Shropshire may, according to Pharaoh et al. (1993), form a significant suite within the Midlands Micro-craton which they referred to as the Midlands Minor Intrusion Suite of Ashgill (Ordovician) age. It has also been suggested that the Barnt Green Volcanic Formation (Tremadoc), which outcrops in the Redditch district, may represent an early extrusive phase to the principal magmatic episode (Carney et al., 1992). The location of magnetic anomalies (Figure 32) close to outcropping rocks of the Barnt Green Volcanic Formation (M4), and to the lamprophyre intrusions of Dost Hill (M7) and Nuneaton (M8), suggest the possibility that some of the anomalies in and around the district could also be attributable to Caledonian (Ordovician) igneous activity.

The basalts and dolerites of Carboniferous age in the area around Rowley Regis [9650 8900] were surveyed geophysically by Marshall (1945) and found to be highly magnetic. However, these rocks only produce a relatively weak aeromagnetic anomaly (approximately 60 nT observed only on flight line records), which, in Figure 32, is largely obscured by the higher-amplitude, longer-wavelength anomaly it is superimposed upon. Clearly, highly magnetic basalts and dolerites as seen at Rowley Regis would produce large magnetic anomalies if they were to occur in sufficient volume.

The magnetic anomalies seen along the profile in Figure 33 are of large amplitude and long wavelength, indicating that they must be due to large volume sources and have therefore been attributed to magnetic material within the Precambrian basement.

In the absence of solid evidence within the district the precise nature of the magnetic material remains uncertain, but it does seem most likely that Uriconian rocks contribute to the anomalies.

INFORMATION SOURCES

Further geological information held by the British Geological Survey relevant to the district is listed below. It includes published maps, memoirs and reports, and open-file maps and reports. Other sources include borehole records, mine plans, fossils, rock samples, thin sections, hydrogeological data and photographs.

Searches of indexes to some of the collections can be made on the Geoscience Index system in British Geological Survey libraries. This is a developing computer-based system which carries out searches of indexes to collections and digital databases for specified geographical areas. It is based on a geographic information system linked to a relational database management system. Results of the searches are displayed on maps on the screen. At the present time (1999) not all of the data sets are complete. The available indexes are listed below.

- Index of boreholes
- Topographical backdrop based on 1:250 000 scale maps
- Outlines of BGS maps at 1:50 000, 1:10 000, 1:10 560 and County Series maps
- Chronostratigraphical boundaries and areas from British Geological Survey 1:250 000 maps
- Geochemical sample locations on land
- Aeromagnetic and gravity data recording stations
- Land survey records

BGS PUBLICATIONS RELEVANT TO THIS DISTRICT

Books

British Regional Geology
Central England, 3rd edition, 1969

Memoirs
The geology of the country around Birmingham (Sheet 168), 1st edition (1925)*
Dudley and Bridgnorth, (Sheet 167, reprinted 1987), 1947
Geology of the country around Redditch (Sheet 183), 1991
Geology of the country around Coventry and Nuneaton (Sheet 169), 1998
Geology of the country around Lichfield (Sheet 154), 1919*
Geology of the southern part of the South Staffordshire Coalfield, (1927)*

* out of print

Maps

1:1 500 000
Colour shaded relief gravity anomaly map of Britain, Ireland and adjacent areas (1996). Smith, I F, and Edwards, J W F (compilers), British Geological Survey
Colour shaded relief magnetic anomaly map of Britain, Ireland and adjacent areas. Royles, C P, and Smith, I F (compilers), British Geological Survey
Metallogenic map of Britain and Ireland, 1996
Tectonic map of Britain, Ireland and adjacent areas, 1996

1:1 000 000
Industrial mineral resources map of Britain, 1996
Pre-Permian geology, Solid (pre-Quaternary) geology (south sheet)

1:625 000
Geological map of the United Kingdom, South, 1979
Quaternary map of the United Kingdom, South, 1977

1:250 000
52N 04W Mid-Wales and Marches, Solid Geology, 1990
52N 02W East Midlands, Solid Geology, 1983
52N 02W Aeromagnetic anomaly, East Midlands, 1980
52N 02W Bouguer gravity anomaly, East Midlands, 1982

1:50 000 and 1:63 360
Sheet 153 Wolverhampton (Solid and Drift) 1993
Sheet 154 Lichfield (Solid and Drift) 1922
Sheet 155 Coalville (Solid and Drift) 1982
Sheet 167 Dudley (Solid and Drift) 1975
Sheet 168 Birmingham (Solid and Drift) 1996
Sheet 169 Coventry (Solid and Drift) 1994
Sheet 182 Droitwich (Solid and Drift) 1976
Sheet 183 Redditch (Solid and Drift) 1989
Sheet 184 Warwick (Solid and Drift) 1984

1: 25 000 and 1:50 000
Thematic (applied) geological maps of parts of the Black Country (15 maps) and Coventry districts (14 maps) accompanying BGS Technical Reports WA/92/33 and WA/89/29, respectively.

1:50 000 Geophysical Information Maps
Plot-on-demand maps are available which summarise graphically the publicly available geophysical information held for the sheets in the BGS databases. Features include:
- Regional gravity data: Bouguer anomaly contours and location of observations.
- Regional aeromagnetic data: total field anomaly contours and location of digitised data points along flight lines.
- Gravity and magnetic fields plotted on the same base map at 1:50 000 scale to show correlation between anomalies.
- Separate colour contour plots of gravity and magnetic fields at 1:125 000 scale for easy visualisation of important anomalies.
- Location of local geophysical surveys.
- Location of public domain seismic reflection and refraction surveys.
- Location of deep boreholes and those with geophysical logs.

1:10 000 and 1:10 560 scale geological maps
BGS 1:10 000 or 1:10 560 scale geological maps included wholly, or in part, in Sheet 168 are listed below, together with the initials of the surveyor and the date of survey and revision survey where applicable. The surveyors were B W Glover, R J O Hamblin, M R Henson, R A Old, D P Piper, J H Powell, M G Sumbler and C N Waters.

Copies of these maps are available for public reference in the libraries of the British Geological Survey in Keyworth and Edinburgh. Uncoloured dyeline copies are available for purchase from BGS; some sheets are available in a digital format.

SO 97 NE	Rubery	RJOH	1980
SO 98 NE	Rowley Regis	CNW	1990
SO 98 SE	Halesowen	BWG	1990

SO 99 NE	Willenhall and Darlaston	RJOH,	1979,
		MRH, JHP	1991
SP 07 NW	Longbridge	RJOH	1979, 1981
SP 07 NE	Hollywood	RAO	1982
SP 08 NW	Smethwick	JHP	1990
SP 08 NE	Birmingham City	JHP	1993
SP 08 SW	Harborne	BWG	1989
SP 08 SE	Edgbaston	BWG	1993
SP 09 NW	Walsall	RJOH,	1978,
		MRH, CNW	1991
SP 09 NE	Streetly	CNW	1993
SP 09 SW	Hamstead	RJOH,	1978,
		MRH, CNW	1991
SP 09 SE	Perry Barr	CNW, JHP	1993
SP 17 NW	Shirley	RAO	1981
SP 17 NE	Solihull and Knowle	RAO	1980
SP 18 NW	Stechford	DPP	1992
SP 18 NE	Chelmsley Wood	JHP	1992
SP 18 SW	Acock's Green	DPP	1993
SP 18 SE	Elmdon	MGS	1982
SP 19 NW	Sutton Coldfield	BWG	1992
SP 19 NE	Middleton	BWG	1992
SP 19 SW	Erdington	DPP	1993
SP 19 SE	Water Orton	CNW	1992
SP 27 NW	Berkswell and Balsall Common	RJOH	1978–80
SP 28 NW	Maxstoke	RAO	1988
SP 28 SW	Meriden	MGS	1982
SP 29 NW	Dost Hill	JHP, BWG	1996
SP 29 SW	Nether Whitacre	JHP	1992

BGS Technical (Open-file) and other reports

GEOLOGY

The Technical Reports and Open-file Reports (marked *) listed below are detailed accounts of the geology of the constituent 1:10 000 scale maps of the Birmingham 1:50 000 Series Sheet 168. Copies of the reports may be ordered from British Geological Survey, Keyworth, Nottingham.

*SO 97 NE   HAMBLIN, R J O. 1984.   *Geological notes and local details for 1:10 000 sheets SO 97 NE (Rubery).*   (Keyworth, Nottingham: Institute of Geological Sciences.)

SO 98 NE   WATERS, C N. 1991.   Geology of the Rowley Regis district.   *British Geological Survey Technical Report*, WA/91/55.

SO 98 SE   GLOVER, B W. 1990a.   Geology of the Halesowen district.   *British Geological Survey Technical Report*, WA/90/74.

SO 99 NE   HAMBLIN, R J O, HENSON, M R, and POWELL J H. 1991.   Geological notes and local details for 1:10 000 sheet SP09NW (Walsall). 2nd edition.   *British Geological Survey Technical Report*, WA/91/63.

SO 99 SE   HAMBLIN, R J O, and GLOVER, B W. 1991.   Geological notes and local details for 1:10 000 sheet SO99SE (Dudley and Wednesbury). 2nd edition.   *British Geological Survey Technical Report*, WA/91/73.

*SP 07 NW   HAMBLIN, R J O. 1984.   *Geological notes and local details for 1:10 000 sheets SP 07NW (Longbridge).*   (Keyworth, Nottingham: Institute of Geological Sciences.)

*SP 07 NE   OLD, R A. 1983.   *Geological notes and local details for 1:10 000 sheets SP 07NE (Hollywood).*   (Keyworth, Nottingham: Institute of Geological Sciences.)

SP 08 NW   POWELL, J H. 1991.   Geology of the Smethwick district (SP08NW).   *British Geological Survey Technical Report*, WA/91/71.

SP 08 NE   POWELL, J H. 1993.   Geology of the Birmingham City district (SP08NE). *British Geological Survey Technical Report*, WA/93/76.

SP 08 SW   GLOVER, B W. 1990b.   Geology of the Harborne district (SP08SW).   *British Geological Survey Technical Report*, WA/90/75.

SP 08 SE   GLOVER, B W. 1993.   Geology of the Edgbaston district (SP08SE). *British Geological Survey Technical Report*, WA/93/82.

SP 09 NW   HAMBLIN, R J O, HENSON, M R, and WATERS, C N. 1991.   Geological notes and local details for 1:10 000 sheet SP09NW (Walsall).   2nd edition.   *British Geological Survey Technical Report*, WA/91/63.

SP 09 NE   WATERS, C N. 1993.   Geology of the Streetly district (SP09 NE).   *British Geological Survey Technical Report*, WA/93/02.

SP 09 SW   WATERS, C N. 1991.   Geology of the Hamstead district (SP09SW).   *British Geological Survey Technical Report*, WA/91/08.

SP 09 SE   WATERS, C N, and POWELL J H. 1993.   Geology of the Perry Barr district (SP09SE).   *British Geological Survey Technical Report*, WA/94/16.

*SP 17 NW   OLD, R A. 1982.   *Geological notes and local details for 1:10 000 sheets SP 17NW (Shirley).*   (Keyworth, Nottingham: Institute of Geological Sciences.)

*SP 17 NE OLD, R A. 1982.   *Geological notes and local details for 1:10 000 sheets SP 17NE (Solihull and Knowle).*   (Keyworth, Nottingham: Institute of Geological Sciences.)

SP 18 NW   PIPER, D P. 1993.   Geology of the Stechford district (SP18NW).   *British Geological Survey Technical Report*, WA/93/52.

SP 18 NE   POWELL, J H. 1993.   Geology of the Chelmsley Wood district (SP18NE).   *British Geological Survey Technical Report*, WA/93/08.

SP 18 SW   PIPER, D P. 1993.   Geology of the Acock's Green district (SP18SW).   *British Geological Survey Technical Report*, WA/93/79.

*SP 18 SE   SUMBLER, M G. 1982.   *Geological notes and local details for 1:10 000 sheets SP 18SE (Elmdon).*   (Keyworth, Nottingham: Institute of Geological Sciences.)

SP 19 NW   GLOVER, B W. 1993.   Geology of the Sutton Coldfield district (SP19NW).   *British Geological Survey Technical Report*, WA/93/32.

SP 19 NE   GLOVER, B W. 1992.   Geology of the Middleton district (SP19NE).   *British Geological Survey Technical Report*, WA/92/67.

SP 19 SW   PIPER, D P. 1993.   Geology of the Erdington district (SP19SW).   *British Geological Survey Technical Report*, WA/93/09.

SP 19 SE   WATERS, C N. 1992.   Geology of the Water Orton district (SP19SE).   *British Geological Survey Technical Report*, WA/92/68.

*SP 27 NW   OLD, R A. 1988.   *Geological notes and local details for 1:10 000 sheets SP 27NW (Berkswell and Balsall Heath).*   (Keyworth, Nottingham: Institute of Geological Sciences.)

SP 28 NW   OLD, R A. 1988.   *Geological notes and local details for 1:10 000 sheets SP 28NW (Maxstoke).   British Geological Survey Technical Report*, WA/89/20.

*SP 28 SW   SUMBLER, M G. 1982.   *Geological notes and local details for 1:10 000 sheets SP 28SW (Meriden).* (Keyworth, Nottingham: Institute of Geological Sciences.)

SP 29 SW   POWELL, J H. 1993.   Geology of the Nether Whitacre district (SP 29SW).   *British Geological Survey Technical Report*, WA/93/08.

SAND AND GRAVEL RESOURCES

Part of the south-east of the district is covered by the following BGS Mineral Assessment Reports (sand and gravel):

CANNELL, B. 1982.   The sand and gravel resources of the country east of Solihull, Warwickshire. Description of 1:25 000 resource sheet SP 17, 18, 27, and 28.   *Mineral Assessment Report of the Institute of Geological Sciences*, No. 115.

CANNELL, B, and CROFTS, R G. 1984.   The sand and gravel resources of the country around Henley-in-Arden, Warwickshire. Description of 1:25 000 resource sheet SP16 and parts of SP 17, 25, 26, and 27.   *Mineral Assessment Report of the British Geological Survey*, No. 142.

In addition, there are two open-file reports dealing with the sand and gravel resources of the district:

OLD, R A. 1982.   *Quaternary deposits of sheets SP 17 and SP27W (Solihull and Balsall Common).*   (Keyworth, Nottingham: Institute of Geological Sciences.)

OLD, R A. 1983.   *Sheets SO 97E and SP 07 (Bromsgrove and Alvechurch). Geology with special emphasis on potential resources of sand and gravel.*   Geological reports for DoE: Land use planning. (Keyworth, Nottingham: Institute of Geological Sciences).

GEOLOGY AND LAND-USE PLANNING

Parts of the district are also covered by the following BGS Technical Reports and accompanying thematic geological maps dealing with land-use planning and development:

Western part of the district (west of Easting 05): POWELL, J H, GLOVER, B W, and WATERS C N. 1992.   A geological background for planning and development in the 'Black Country'.   *British Geological Survey Technical Report*, WA/92/33.

South-eastern part of the district (SP 27 NW, 28 SW and 28 NW): OLD, R A, BRIDGE, McC D, and REES, J G. 1990.   Geology of the Coventry area.   *British Geological Survey Technical Report*, WA/89/29.

ENGINEERING GEOLOGY

FORSTER, A. 1991.   The engineering geology of Birmingham West (the Black Country).   *British Geological Survey, Technical Report*, WN/91/15.

GEOPHYSICS

ROYLES, C P. 1996.   Geophysical investigations in the Birmingham district. 1995. *British Geological Survey Technical Report*, WK/96/10.

BIOSTRATIGRAPHY

MOLYNEUX, S G. 1987.   Palynology report: Microfossils from the Lickey Quartzite.   *Unpublished palaeontological report* PD/87/382. British Geological Survey.

RUSHTON, A W A. 1990.   Note on a trilobite collected from the Enville Formation.   *British Geological Survey Technical Report, Stratigraphy Series*, WA/90/234R.

RUSHTON, A W A. 1993.   Report on the basement beds in the Roughs Borehole.   *British Geological Survey Technical Report, Stratigraphy Series*, WA/93/1C.

RUSHTON, A W A. 1994.   Fossils from a clast in the Hopwas Breccia, Great Barr, Birmingham.   *British Geological Survey Technical Report, Stratigraphy Series*, WA/94/183R.

RUSHTON, A W A. 1994.   Fossils from the Rubery Sandstone near Shustoke Farm, Great Barr, Birmingham.   *British Geological Survey Technical Report, Stratigraphy Series*, WA/94/184R.

RUSHTON, A W A. 1994.   Cambrian fossils from the Nechells Breccia.   *British Geological Survey Technical Report, Stratigraphy Series*, WA/94/287R.

RUSHTON, A W A. 1994.   Review of selected boreholes reaching Lower Palaeozoic rocks around Birmingham.   *British Geological Survey Technical Report, Stratigraphy Series*, WA/94/297R.

RUSHTON, A W A. 1995.   Illustration of Cambrian and Ordovician fossils from the Birmingham district.   *British Geological Survey Technical Report, Stratigraphy Series*, WA/95/96R.

RUSHTON, A W A, and TUNNICLIFF, S P. 1994.   Fossils from the basement beds in the Shenstone borehole.   *British Geological Survey Technical Report, Stratigraphy Series*, WA/94/184R.

WARRINGTON, G. 1972.   Palynology report: Arden Sandstone, Hampton-in-Arden, (Sheet 168).   *Unpublished palaeontological report* PD/80/123. British Geological Survey

WARRINGTON, G. 1993.   Palynology report: Mercia Mudstone Group (Triassic), Holly Lane brickpit, Erdington, Birmingham.   *British Geological Survey Technical Report, Stratigraphy Series*, WH/93/188R.

WARRINGTON, G. 1993.   Palynology report: Mercia Mudstone Group (Triassic), Jackson's brickpit, Stonebridge, near Hampton in Arden.   *British Geological Survey Technical Report, Stratigraphy Series*, WH/93/189R.

WARRINGTON, G. 1993.   Palynology report: Arden Sandstone Formation, Mercia Mudstone Group (Triassic); borehole '102R', Middle Bickenhill, near Hampton in Arden.   *British Geological Survey Technical Report, Stratigraphy Series*, WH/93/194R.

WARRINGTON, G. 1993.   Palynology report: Arden Sandstone Formation, Mercia Mudstone Group (Triassic); borehole '101R', Stonebridge, near Hampton in Arden.   *British Geological Survey Technical Report, Stratigraphy Series*, WH/93/195R.

WARRINGTON, G. 1993.   Palynology report: Arden Sandstone Formation, Mercia Mudstone Group (Triassic); borehole '105R', Stonebridge, near Hampton in Arden.   *British Geological Survey Technical Report, Stratigraphy Series*, WH/93/196R.

SEDIMENTOLOGY

HALLSWORTH, C R. 1992a.   Stratigraphic variations in the heavy minerals and clay mineralogy of the Westphalian to ?Early Permian succession from the Daleswood Farm borehole, and the implications for sand provenance.   *British Geological Survey Technical Report*, WH/92/185R.

HALLSWORTH, C R. 1992b.   Stratigraphic variations in the heavy minerals of the Westphalian succession from the Romsley Borehole.   *British Geological Survey Technical Report*, WH/92/186R.

HALLSWORTH, C R. 1994.   Heavy mineral characterisation of Carboniferous to Triassic sandstones in the Warwickshire area and their importance for provenance.   *British Geological Survey Technical Report*, WH/94/88/R

JONES, N S. 1992.   Sedimentology of the Langsettian (Westphalian A) and Duckmantian (Westphalian B) from the Coventry area of the Warwickshire Coalfield.   *British Geological Survey Technical Report*, WH/92/172C.

LOTT, G K. 1992a.   Petrology and diagenesis of the Upper Carboniferous sandstones from the Daleswood Farm borehole, South Staffordshire.   *British Geological Survey Technical Report*, WH/92/182R.

LOTT, G K. 1992b.   Petrology, diagenesis and provenance of the Upper Carboniferous (Westphalian) sandstones from the South Staffordshire area.   *British Geological Survey Technical Report*, WH/92/198R.

LOTT, G K. 1993.   Thin section petrography of Upper Carboniferous (Westphalian) sandstones from the South Staffordshire area.   *British Geological Survey Technical Report*, WH/93/122R.

BOREHOLES AND SHAFTS

Boreholes mentioned in the text and selected boreholes in and adjacent to the district are listed alphabetically below, together with their National Grid reference, BGS registered number (1:10 000 scale quarter-sheet number), and total depth; ug indicates underground borehole.

Borehole data for the district are catalogued in the BGS archives (National Geoscience Records Centre) at Keyworth on individual 1:10 000 scale sheets, each catalogue consisting of a site map at 1:10 560 or 1:10 000 scale and a borehole register, together with the individual records. For further information contact: The Manager, National Geosciences Records Centre, British Geological Survey, Keyworth, Nottingham NG12 5GG.

Borehole name	National Grid reference	BGS registered number	Total depth (m)
Aston Wells Pumping Station	SP 0918 9034	SP 09 SE 8	127
Ansells Brewery No. 5	SP 0807 8914	SP 08 NE 13	162
Berryfields Farm (NCB)	SP 2499 8148	SP 28 SW 179	1012
Birmingham Race Course	SP 1282 8970	SP 18 NW 91	306
Blackham Colliery No. 1 Pit	SO 9864 9627	SO 99 NE 393	163
Blakeley Hall Colliery No. 2 Pit	SO 9992 8952	SO 98 NE 42b	288
Blind Lane (NCB)	SP 2450 7962	SP 27 NW 2	1043
Blyth Bridge (NCB)	SP 2119 8979	SP 28 NW 11	1048
Bradley Colliery, No. 15	SO 9602 9485	SO 99 SE 4	151
Bromford Tube Co. (a)	SP 1115 9005	SP 19 SW 73	212
Bromford Tube Co. (b)	SP 1131 8985	SP 18 NW 195	260
Bromford Lane	SP 1176 8919	SP 18 NW 80	92
Coombeswood Colliery	SO 9715 8461	SO 98 SE 4	247
Cow Pasture Pits	SO 9847 9551	SO 99 NE 938	47
Daleswood Farm (BGS)	SO 9512 7913	SO 97 NE 452	256
Dixon and Burne Colliery No. 212	SO 9672 9158	SO 99SE 3187	218
Dove House Farm	SP 2472 8912	SP 28 NW 1	643
Dumble Farm	SP 2306 8874	SP 28 NW 6	275
Dunlop Rubber (Hercules)	SP 0928 8983	SP 08 NE 7	107
Dunlop Rubber (Fort Dunlop)	SP 1225 9017	SP 19 SW 66	270
Flanders Hall	SP2311 9429	SP 29 SW 3	336
F Smith Brewery	SP 0871 8945	SP08NE 4	122
Guest Keen and Nettlefolds No. 2	SP 0342 8855	SP 08 NW 82	135
Haden Hill Colliery	SO 9604 8573	SO 98 NE 34	224
Hamstead No. 1	SP 0760 9625	SP 09 NE 7	949
Hamstead No. 2 (Great Barr)	SP 0503 9566	SP 09 NE 1	601
Hamstead Colliery No.1 Shaft	SP 0431 9296	SP 09 SW 34	597
Heath Pits	SP 0071 9117	SP 09 SW 27	282
Horseley New Colliery No. 3	SO 9697 9223	SO 99 SE 472	224
HP Sauce (Midland Vinegar)	SO 0794 8910	SP 08 NE 347	217
Hunnington	SO 965 815	SO 98 NE 90	120
J & E Sturge Ltd No. 2	SP 0579 7960	SP 07 NE 197	370
J & E Sturge Ltd No. 1 (Kings Norton)	SP 0587 7963	SP 07 NE 23	366
Kate's Hill No. 2	SO 9508 8994	SO98 NE 90	43
Kimberley's Grove (NCB)	SP 2462 8763	SP 28 NW 7	671

Borehole name	National Grid reference	BGS registered number	Total depth (m)
Kingsbury Borehole	SP 2446 9449	SP 29 SW 1	314
Kingsbury No. 2 Underground	SP 2470 9374	SP29 SW 2	49 ug
Kingsbury No. 3 Underground	SP 2402 9354	SP 29 SW 4	92 ug
Kitt's Green (J Booth's)	SP 1481 8755	SP 18 NW 192	413
Knowle Borehole (IGS)	SP 1883 7777	SP 17 NE 184	132
Longbridge Laundry	SP 0201 7847	SP 07 NW 45	118
Longbridge Pumping Station	SP 0072 7755	SP 07 NW 3	151
Lucas & Co. (Works)	SP 1029 8355	SP 18 SW 35	305
Manor Pit	SO 9766 8311	SO 98 SE 5	287
Metro-Cammell–Railway Carriage Works, Washwood Heath	SP 1065 8925	SP 18 NW 1	198
Mitchells and Butlers (City Road)	SP 0366 8688	SP 08 NW 475	203
Moat Farm	SP 0025 8770	SP 08 NW 1	446
Moss Gear Co.	SP 1377 9151	SP 19 SW 69A	221
Mucklow Hill	SO 9766 8469	SO 98 SE 6	177
Moxley Colliery No. 5	SO 9625 9620	SO 99 NE 378	121
Nechells Gasworks No. 1	SP 0910 8874	SP 08 NE 18	219
Nechells Gasworks No. 3	SP 0933 8831	SP 08 NE 322	306
Nechells Redevelopment	SP 0873 8813	SP 08 NE 192	23
Ocker Hill Colliery No.7 Pit	SO 9787 9383	SO 99 SE 537	141
Old Blackheath Colliery	SO 9760 8592	SO 98 NE 57	177
Outwoods (NCB)	SP 2462 8528	SP 28 NW 52	611
Great Packington (Spring Pools)	SP 2471 8479	SP 28 SW 1	612
Priory Wood (NCB)	SP 2361 8578	SP 28 NW 55	406
Quinton No. 1	SO 9921 8471	SO 98 SE 126	31
Railway Carriage Works (Birmingham)	SP 0285 8925	SP 08 NW 5	201
Ram Hall	SP 2469 7809	SP 27 NW 3	1046
Renold & Coventry (Birch Road)	SP 0859 9094	SP 09 SE 482	113
Romsley	SO 9501 7893	SO 97 NE 216	67
Rowley Hall Colliery No. 9 Pit	SO 9752 8756	SO98 NE 50	212
Rubery Hospital	SO 995 780	SO 97 SE	c. 3
Sandwell Park No. 1	SP 0199 8980	SP 08 NW 2	472
Sandwell Park Colliery (Diamond Jubilee Pits No. 4 Shafts)	SP 0257 9225	SP 09 SW 234B	350
Schweppes Beverages	SP 1430 9196	SP 19 SW 99	244
Smart and Son	SP 0809 8674	SP 08 NW 42	222
Southalls No. 4	SP 1080 8778	SP 18NW 78	305
Tipton Green No. 6	SO 9597 9249	SO 99SE 29	171
Tividale Hall Colliery No. 8	SO 9656 9093	SO 99 SE 258	219
Trickley Lodge (NCB)	SP 1603 9884	SP 19 NE 22	876
Vickers Armstrong (Castle Bromwich)	SP 1290 9060	SP 19 SW 51	304
Walbutts Colliery No. 3	SO 9608 9621	SO 99NE 376	106
Whitehall Colliery	SO 9824 9179	SO 99 SE 313	41
Whitehouse Farm Borehole (NCB)	SP 2387 9303	SP 29 SW 8	500
Wilmot Breedon	SP 1185 8446	SP 18 SW 143	304
Windsor Street Gasworks No. 3	SO 0797 8822	SP 08 NE 325	285
Woodcock Wood (NCB)	SP 2425 8681	SP 28 NW 56	678
Walsall Borehole (Co-op Society)	SP 0093 9805	SP 09 NW 33	392

OTHER SOURCES AND TYPES OF INFORMATION

GEOPHYSICS

Gravity and aeromagnetic data are held digitally in the National Gravity Databank and the National Aeromagnetic Databank at BGS Keyworth.

Geophysical information maps are available for Birmingham Sheet 168 and adjacent sheets.

There is very little seismic reflection data for the district. Profiles for the westernmost part of the district and surrounding areas are from surveys made by the National Coal Board, British Coal and Harlech Petroleum.

HYDROGEOLOGY

Data on water boreholes, wells and springs are held in the BGS (Hydrogeology Group) database at Wallingford. Similar data are held by the National Rivers Authority, and consist of:

- A data bank of all licensed wells, boreholes and springs. These data give the permitted rates of abstraction, the specific location and the depth and diameter of the well or borehole. Abstraction data are confidential where they relate to individual industrial users.
- A data bank of groundwater chemistry relating to all public water supply sources and a few selected industrial sources.

• Limited data on long-term changes in groundwater levels.

MINERALS

Directory of Mines and Quarries

United Kingdom Minerals Yearbook

MINGOL is a GIS-based minerals information system, from which hard-copy and digital products tailored to individual clients' requirements can be obtained.

THIN SECTIONS

Thin sections of rocks from the district are held in the England and Wales Sliced Rocks collection at BGS Keyworth. Charges and conditions of access to the Collection are available on request from BGS Keyworth.

FOSSILS

Macrofossils and micropalaeontological residues for samples collected from the district are held at BGS Keyworth. Enquiries concerning all macrofossil material should be directed to the Curator, Biostratigraphy Collections, BGS Keyworth.

BGS STRATIGRAPHICAL LEXICON

Definitions of the named rock units shown on BGS maps, including those shown on the 1:50 000 Series Birmingham Sheet 168 are held in the BGS Lexicon of Named Rock Units. The Lexicon can be accessed at Website: http://www.bgs.ac.uk, under 'Free products' Further information on the database can be obtained from the Lexicon Manager at BGS Keyworth.

BGS PETMIN DATABASE

A database of thin sections and rock samples is maintained by the Mineralogy and Petrology Group at BGS Keyworth. The Group Manager at Keyworth should be contacted for further information, including methods of accessing the database.

BGS (GEOLOGICAL SURVEY) PHOTOGRAPHS

Copies of these photographs are deposited for reference in the British Geological Survey Library, Keyworth, Nottingham NG12 5GG. Colour or black and white prints and transparencies can be supplied at a fixed tariff.

The National Grid references are those of the viewpoints, where known.

Carboniferous and Permian rocks

(i)	**Late Carboniferous and Early Permian strata**
1551	Sandstone with conglomerate bands in Enville Member. Canal at Hamstead, near Great Barr.
1552	Sandstone in Enville Member, canal at Hamstead, near Great Barr.
1553–5	Red mudstone with sandstone bands in Anville Member. Large Quarry, Hamstead.
1950–1	Conglomerate rich in Carboniferous limestone clasts in Enville Member. Small quarry south-west of Bristnall Fields, near Oldbury.
1990	'Espley' sandstone in Etruria Formation. Road cutting, Coombes Road, near Halesowen Station.
2005	Clent Formation, showing dip. Lane, about 800 m north-west of the Bell Inn, Northfield.

2011–5	Panoramic views, showing the type of scenery produced by the Clent Formation. From the Lickey Hills, Rednall.
2016	Outlier of the Clent Formation. Frankley Upper Beeches, from Ley Hill, one mile east of Frankley.
2024–5	Sandstone in lower part of Halesowen Formation. Road-cutting, Bromsgrove Road, about 400 m south-west of Halesowen Station.

(ii)	**Dolerite**
1937	Rowley Regis dolerite intruded into Etruria Formation. Hailstone Quarry, Rowley Regis.
1938–9	Joints in Rowley Regis dolerite. Hailstone Quarry, Rowley Regis.
1940–1	Crusher, elevator, and screens, respectively. Hailstone Quarry, Rowley Regis.
1943–4	Curved joint surfaces in dolerite. Central Quarry, Tippity Green, Rowley Regis.
1945	General view of quarry in dolerite. Prospect Quarry, Rowley Regis.
1946–7	Columnar jointing in dolerite. Prospect Quarry, Rowley Regis.
1948	Crushed band in dolerite. Prospect Quarry, Rowley Regis.
1949	Rock drill. Prospect Quarry, Rowley Regis.

Triassic rocks

1557	Hopwas Breccia overlain by Kidderminster Formation conglomerate. Quarry near Blackroot Pool, Sutton Park.
1558	Hopwas Breccia with sandstone beds. Quarry near Blackroot Pool, Sutton Park.
1559	Pebbly sandstone of the Hopwas Breccia. Disused quarry on roadside, 300 m south of Barr Beacon, near Great Barr.
1993	Wildmoor Sandstone overlain by drift. Cemetery sand pit, Key Hill, Hockley, Birmingham.
1994	Mercia Mudstone with pale 'skerry' bands. Hough and Co.'s Brickworks, 800 m south of Hazelwell Station.
2001	Mercia Mudstone, nearly horizontal, overlain by till and glaciofluvial sand and gravel. King's Norton Brickworks.
2002	Pocket of glaciofluvial gravel in Mercia Mudstone. King's Norton Brickworks.
2003	Mercia Mudstone ('Waterstones' facies) overlain by till. Pigeon House Hill, Northfield.
2004	Mercia Mudstone ('Waterstones' facies), cross-bedded and faulted. Road cutting, Pigeon House Hill, Northfield.
13523–4	Mercia Mudstone. Jackson's Brick Quarry, Middle Bickenhill.

Quaternary deposits

1549–50	Glaciofluvial and glaciolacustrine deposits infilling an old channel in Coal Measures, Moxley Sand Pit, near Darlaston.
1556	Contorted glaciofluvial sand and gravel. Bustleholm Sand Pit. 1.6 km north-west of Newton Road Station, West Bromwich.
1987–9	Highly contorted clay, sand and gravel (glaciofluvial and glaciolacustrine deposits). Bustleholm Sand Pit.
1991–2	Glaciolacustrine clay with fine-grained sand and gravel. Clay pit, California Brickworks, near Harborne.

1995	Section in glaciofluvial sand and gravel (Mosley gravels). Gravel pit, Northlands Road, King's Heath.
1996	Section in glaciofluvial sand and gravel (Mosley gravels). Gravel pit, Wake Green Road, Springfield.
1997–8	Section in glaciofluvial sand and gravel (Mosley gravels). Gravel pit, Swanshurst Lane, 800 m south of Springfield.
1999	Section in glaciofluvial sand and gravel. Little's Sand Pit, Brook Lane, Billesley Common.
2000	Contorted glaciofluvial sand and gravel. Little's Sand Pit, Brook Lane, Billesley Common.
13518	The Somers Gravel Pit, Meriden [SP 225 821].
13519	Cornet's End Gravel Pit, Meriden [SP 233 814].
13520	Glaciolacustrine sands and glaciofluvial gravels. Cornet's End Gravel Pit [SP 236 815].
13521	Glaciolacustrine sands. Cornet's End Gravel Pit, Meriden [SP 236 815].
13522	Till lens in glaciolacustrine sand. Cornet's End Gravel Pit, Meriden [SP 236 815].
13765–9	Glaciofluvial sand and gravel. Shirley Quarry [SP 096 779].
13770	Glaciolacustrine clay. Shirley Quarry [SP 0986 7810].
13771–2	Glaciofluvial sand and gravel. Shirley Quarry [SP 0980 7818 and SP 0960 7650].

LIST OF MINERAL INDUSTRY OPERATORS (1994)

Listings, by county, refer to the site name, National Grid Reference, operator and geological unit (Harris et al., 1994). Some of the workings may have ceased at the time of publication of this memoir.

1. Sand and gravel

Warwickshire

Dunton, Coleshill [SP 188 933], Landfill Developments, glacial and glaciolacustrine sand and gravel.
Middleton Hall, Sutton Coldfield [SP 195 975], ARC — Central, alluvium of the River Tame.
Blyth Hall, Coleshill [SP 205 905], RMC — Western Aggregates, alluvium of the River Tame.

2. Common clay (brick and tile clay)

Warwickshire

Arden Works, Hampton-in-Arden [SP 205 830], Packington Estates Enterprises Ltd. (formerly Redland Bricks site), Mercia Mudstone Group.

3. Dolerite aggregate

West Midlands

Edwin Richards, Dudley [SO 969 883], ARC — Central, dolerite (Carboniferous).

ADDRESSES FOR DATA SOURCES AND SUMMARY OF MAIN SERVICES AND PRODUCTS AVAILABLE

BGS ENQUIRY SERVICE
1:10 000 maps (sales and reference copies); borehole, shaft and trial pit records; borehole samples; geophysical data; remote sensing data; geochemical data; fossils; thin sections; Petmin database; Stratigraphical Lexicon; library; publications sales desk.

British Geological Survey, Headquarters, Sir Kingsley Dunham Centre, Keyworth, Nottingham NG12 5GG.
Telephone 0115-936 3100. Fax 0115-936 3200
Web site: http://www.bgs.ac.uk

BGS HYDROGEOLOGY ENQUIRY SERVICE
Wells, springs and water borehole records.

British Geological Survey, Hydrogeology Group, Maclean Building, Crowmarsh Gifford, Wallingford, Oxfordshire OXO 8BB. Telephone 01491-838800. Fax 01491-692345.

GROUNDWATER; FLOOD RISK; LANDFILL SITES.

Environment Agency, Severn-Trent Region, Saphire East, 550, Streetsbrook Road, Solihull, B91 1QT. Telephone 0121-711 2324. Fax 0121-711 5824.

MINE PLANS

Coal, ironstone and fireclay
Copies of all known colliery abandonment plans, together with many of the abandonment plans of the Silurian limestone mines, are held by the Mining Records Office, Coal Authority, Bretby Business Park, Ashby Road, Burton on Trent, Staffordshire DE15 0QD. Mine abandonment plans are held by the Coal Authority in the public domain; they are not available for reference at BGS.

Silurian limestone
In addition to the abandonment plans of Silurian limestone mines held by the Coal Authority, noted above, an up-to-date register of all the known limestone mines and approximate outlines of the workings are held by the Building Control Officer of the relevant Metropolitan District Council, or City Council.

GEOLOGICAL CONSERVATION

Geological conservation is administered by English Nature, Northminster House, Peterborough PE1 1UA. Telephone 01733-340345. Additional information may be obtained from:

Black Country Geological Society
Honorary Secretary
16 St Nicolas Gardens
Kings Norton
Birmingham B38 8TW

Keeper of Geology
Dudley Museum and Art Gallery
St James' Road
Dudley DY1 1HU

At the time of writing, a single geological Site of Special Scientific Interest (SSSI) (Chapter Two) and a number of Regionally Important Geological/Geomorphological Sites (RIGS) are located in the district; up-to-date listings can be obtained from the addresses listed above.

REFERENCES

Most of the references listed below are held in the Library of the British Geological Survey at Keyworth, Nottingham. Copies of the references can be purchased subject to current copyright legislation.

AITKENHEAD, N, and WILLIAMS, G M. 1987. Geological evidence to the Public Inquiry into the Gas Explosion at Loscoe. *British Geological Survey Report*, FP/87/8/83AS.

ALLEN, J R L. 1965. A review of the origin and characteristics of Recent alluvial sediments. *Sedimentology*, Vol. 5, 89–191

ALLPORT, S. 1870. On the basaltic rocks of the Midland coalfields. *Geological Magazine*, Vol. 7, 159–62.

ALLPORT, S. 1884. On the microscopic structure and composition of British Carboniferous dolerites. *Quarterly Journal of the Geological Society of London*, Vol. 30, 529–567.

ALLSOP, J M. 1981. Geophysical appraisal of some geophysical problems in the English Midlands. *Report of the Deep Geology Unit, Institute of Geological Sciences*, No. 81/7.

ANON. 1981a. Concrete in sulphate-bearing soils and groundwaters. *Building Research Establishment Digest*, No. 250.

ANON. 1981b. British Standard code of practice for site investigations, BS5930. *British Standards Institute*. (London: HMSO.)

ANON. 1990a. *This Common Inheritance; Britain's environmental strategy.* Cm. 1200 Summary, 36pp. and white paper. 295pp. (London: HMSO.)

ANON. 1990b. Methods of test for soils for civil engineering purposes BS1377. *British Standards Institute.* (London: HMSO.)

ARBER, E A N. 1909. On the affinities of the Triassic plant *Yuccites vogesiacus*, Schimper and Mougeot. *Geological Magazine*, Vol. 46, 11–14.

ARTER, G. 1983. Geophysical investigations of the deep geology of the East Midlands. Unpublished PhD thesis, University of Leicester.

ARTHURTON, R S. 1980. Rhythmic sedimentary sequences in the Triassic Keuper Marl (Mercia Mudstone Group) of Cheshire, northwest England. *Geological Journal*, Vol. 15, 43–58.

ASHURST, J, and DIMES, F G (editors). 1990. *Conservation of building stone and decorative stone.* (London: Butterworth-Heinemann.)

AUDLEY-CHARLES, M G. 1970. Triassic palaeogeography of the British Isles. *Quarterly Journal of the Geological Society of London*, Vol. 126, 49–89.

BALL, H W. 1951. The Silurian and Devonian rocks of Turner's Hill and Gornal, South Staffordshire. *Proceedings of the Geologists' Association.* Vol. 62, 225–236.

BALL, H W. 1980. Spirorbis from the Triassic Bromsgrove Sandstone Formation (Sherwood sandstone Group) of Bromsgrove, Worcestershire. *Proceedings of the Geologists' Association.* Vol. 19, 149–154.

BARCLAY, W J, AMBROSE, K, CHADWICK, R A, and PHARAOH, T C. 1997. Geology of the country around Worcester. *Memoir of the British Geological Survey*, Sheet 199.

BARNES, A A. 1927. Cementation of strata below reservoir embankments. *Institute of Water Engineers*, 42–48.

BARROW, G, GIBSON, G, CANTRILL, T C, DIXON, E E L, and CUNNINGTON, C H. 1919. The geology of the country around Lichfield. *Memoir of the Geological Survey of Great Britain*, Sheet 154 (England and Wales).

BASSETT, M G. 1974. Review of the stratigraphy of the Wenlock Series in the Welsh Borderland and South Wales. *Palaeontology*, Vol. 17, 745–777.

BASSETT, M G. 1989. The Wenlock Series in the Wenlock area. 51–73 *in* A global standard for the Silurian system. HOLLAND, C H, and BASSETT, M G (editors). *National Museum of Wales Geological Series*, No. 9.

BASSETT, M G, COCKS, L R M, HOLLAND, C H, RICKARDS, R B, and WARREN, P T. 1975. The type Wenlock Series. *Report of the Institute of Geological Sciences*, No. 75/13, 1–19.

BELL, F G. 1981. *Engineering properties of soils and rocks.* (London: Butterworths.)

BENFIELD, A C, and WARRINGTON, G. 1988. New records of the Westbury Formation (Penarth Group, Rhaetian) in North Yorkshire, England. *Proceedings of the Yorkshire Geological Society*, Vol. 47, 29–32.

BENTON, M J. 1990. The species of Rhynchosaurus, a rhynchosaur (Reptilia, Diapsida) from the Middle Triassic of England. *Philosophical Transactions of the Royal Society of London*, B, Vol. 328, 213–306.

BENTON, M J, WARRINGTON, G, NEWELL, A J, and SPENCER, P S. 1994. A review of the British Middle Triassic tetrapod assemblages. 131–160 in *In the shadow of the dinosaurs: early Mesozoic tetrapods.* FRASER, N C, and SUES, H-D (editors). (Cambridge: Cambridge University Press.)

BENTON, M J, and SPENCER, P S. 1995. British Triassic fossil reptile sites. 33–95 in *Fossil reptiles of Great Britain.* Joint Nature Conservation Committee. (London: Chapman and Hall.)

BESLY, B M. 1987. Sedimentological evidence for Carboniferous and Early Permian palaeoclimates of Europe. *Annales de la Société géologique Nord*, Vol. 16, 131–143.

BESLY, B M. 1988. Palaeogeographic implications of late Westphalian to early Permian red-beds, Central England. 200–221 in *Sedimentation in a synorogenic basin complex: the Upper Carboniferous of Northwest Europe.* BESLY B M, and KELLING G (editors). (Glasgow and London: Blackie.)

BESLY, B M, and CLEAL, C. 1997. Upper Carboniferous stratigraphy of the West Midlands (UK) revised in the light of borehole geophysical logs and detrital compositional suites. *Geological Journal*, Vol. 32, 85–118.

BESLY, B M, and TURNER, P. 1983. Origin of red-beds in a moist tropical climate (Etruria Formation). 131–147 in Residual Deposits. WILSON, R C L (editor). *Special Publication of the Geological Society of London*, No. 11.

BESLY, B M, and FIELDING, C R. 1989. Palaeosols in Westphalian coal-bearing and red-bed sequences, Central and Northern England. *Palaeogeography, Palaeoclimatology, Palaeoecology*, Vol. 70, 303–330.

BOSWELL, P G H. 1919. Moulding sands for non-ferrous foundry work. *Journal of the Institute of Metals*, Vol. 22, 277.

BOULTON, G S, and HINDMARSH, R C A. 1987. Sediment deformation beneath glaciers: rheology and geological consequences. *Journal of Geophysical Research*, Vol. 92 (B9), 9059–9082.

BOULTON, W S. 1924. On a recently discovered breccia bed underlying Nechells (Birmingham) and its relation to the red rocks of the district. *Quarterly Journal of the Geological Society of London*, Vol. 80, 343–373.

BOULTON, W S. 1933. The rocks between the Carboniferous and the Trias in the Birmingham district. *Quarterly Journal of the Geological Society of London*, Vol. 84, 53–82.

BOWEN, D Q. 1994. The Pleistocene of North West Europe. *Science Progress Oxford*, Vol. 76, 209–223.

BRASIER, M D. 1984. Microfossils and small shelly fossils from the Lower Cambrian *Hyolithes* Limestone at Nuneaton, English Midlands. *Geological Magazine*, Vol. 121, 229–253.

BRASIER, M D. 1986. The succession of small shelly fossils (especially conoidal microfossils) from English Precambrian–Cambrian boundary beds. *Geological Magazine*, Vol. 123, 237–256.

BRASIER, M D. 1992. Southern British Isles: Comley. 13–14 *in* Atlas of palaeogeography and lithofacies. COPE, J C W, INGHAM, J K, and RAWSON, P F (editors). *Memoir of the Geological Society of London*, No. 13.

BRIDGE, D M, CARNEY, J N, LAWLEY, R S, and RUSHTON, A W A. 1998. Geology of the country around Coventry and Nuneaton. *Memoir of the British Geological Survey*, Sheet 169.

BRIDGLAND, D R, KEEN, D H, and MADDY, D. 1989. The Avon terraces: Cropthorne, Ailstone and Eckington. 51–67 *in* The Pleistocene of the West Midlands: Field Guide. KEEN, D H (editor). (Cambridge: Quaternary Research Association.)

BRITISH GEOLOGICAL SURVEY. 1994. Coventry. England and Wales Sheet 169. Solid and Drift. 1:50 000. (Southampton: Ordnance Survey for British Geological Survey.)

BRITISH GEOLOGICAL SURVEY. 1996. Birmingham. England and Wales Sheet 169. Solid and drift. 1:50 000. (Southampton: Ordnance Survey for British Geological Survey.)

BRODIE, P B. 1865. On the Lias outliers at Knowle and Wooton Warren in south Warwickshire, and on the presence of the Lias or Rhaetic Bone-bed at Copt Heath, its furthest northern extension hitherto recognised in that county. *Quarterly Journal of the Geological Society of London*, Vol. 21, 159–161.

BRODIE, P B. 1874. Notes on a railway-section of the Lower Lias and Rhaetics between Stratford-on-Avon and Fenny Compton, on the occurrence of the Rhaetics near Kineton, and the insect-beds near Knowle, in Warwickshire, and on the recent discovery of the Rhaetics near Leicester. *Quarterly Journal of the Geological Society of London*, Vol. 30, 746–749.

BROOKS, M. 1968. The geological results of gravity and magnetic surveys in the Malvern Hills and adjacent districts. *Geological Journal*, Vol, 6, 13–30.

BROOKS, M, and FENNING, P J. 1968. *In* Geology of the country around Church Stretton, Craven Arms, Wenlock Edge and Brown Clee. GREIG, D C, WRIGHT, J E, HAINS, B A, and MITCHELL, G H (editors). *Memoir of the Geological Survey of Great Britain*, Sheet 166. (England and Wales.)

BROWN, T A. 1980. The Pleistocene history of the Tame catchment, East Birmingham. Unpublished PhD thesis, University of Birmingham.

BUSBY, J P. 1987. An interactive FORTRAN 77 Program using GKS graphics for 2.5D modelling of gravity and magnetic data. *Computers and Geosciences*, Vol. 13, 639–644.

BUSBY, J P, KIMBELL, G S, and PHARAOH, T C. 1993. Integrated geophysical/geological modelling of the Caledonian and Precambrian basement of southern Britain. *Geological Magazine*. Vol. 130, 593–604.

BUTLER, A J. 1937. On Silurian and Cambrian Rocks encountered in a deep boring at Walsall, South Staffordshire. *Geological Magazine*, Vol. 74, 241–257.

BUTLER, A J. 1939. The stratigraphy of the Wenlock Limestone of Dudley. *Quarterly Journal of the Geological Society of London*, Vol. 95, 37–74.

BUTLER, A J, and LEE, J. 1943. Water supply from underground sources in the Birmingham–Gloucester district. *Geological Survey Wartime Pamphlet*, No. 32

CAMPBELL SMITH, W. 1963. Description of the igneous rocks represented among pebbles from the Bunter Pebble Beds of the Midlands of England. *Bulletin of the British Museum, Mineralogy*, Vol. 2, 1–17.

CANNELL, B. 1982. The sand and gravel resources of the country east of Solihull, Warwickshire. Description of 1:25 000 resource sheet SP 17, 18, 27, and 28. *Mineral Assessment Report of the Institute of Geological Sciences*, No. 115,

CANNELL, B, and CROFTS, R G. 1984. The sand and gravel resources of the country around Henley-in-Arden, Warwickshire. Description of 1:25 000 resource sheet SP16 and parts of SP 17, 25, 26, and 27. *Mineral Assessment Report of the British Geological Survey*, No. 142.

CANTRILL, T C. 1909. *Spirorbis* Limestones in the 'Permian' of the South Staffordshire and Warwickshire coalfields. *Geological Magazine*, Vol. 46, 447–454.

CARNEY, J, GLOVER, B J, and PHARAOH, T C. 1992. Pre-conference field excursion guide: Midlands. *British Geological Survey Technical Report*, WA/92/72.

CHADWICK, R A. 1985. Seismic reflection investigations into the stratigraphy and structural evolution of the Worcester Basin. *Journal of the Geological Society of London*, Vol. 142, 187–202.

CHADWICK R A, and SMITH, N J P. 1988. Evidence of negative structural inversion beneath central England from new seismic reflection data. *Journal of the Geological Society of London*, Vol. 145, 519–522

CHANDLER, R J. 1969. The effect of weathering on the shear strength properties of Keuper Marl. *Geotechnique*, Vol. 19, 321–324.

CHANDLER, R J, BIRCH, N, and DAVIS, A G. 1968. Engineering properties of Keuper Marl. *Construction Industry Research and Information Association, London, Research Report*, No. 13

CHARSLEY, T J, RATHBONE, P A, and LOWE, D J. 1990. Nottingham: A geological background to planning and development. *British Geological Survey Technical Report*, WA/90/1.

CHILCOTT, B G. 1922. On the sand and gravel deposits (Glacial Drift) at Stone Cross near West Bromwich. *Proceedings of the Birmingham Natural History and Philosophical Society*, Vol. 15, 19–21.

CIRIA. 1973. Further work on the engineering properties of Keuper Marl. *Construction Industry Research and Information Association, London, Research Report*, No. 47

CLAYTON, G, COQUEL, R, DOUBINGER, J, GUEINN, K J LOBOZIAK, S, OWENS, B, and STREEL, M. 1977. Carboniferous miospores of Western Europe: illustration and zonation. *Mededelingen Rijks Geologische Dienst*, No. 29, 1–71.

COBBOLD, E S. 1921. The Cambrian horizons of Comley (Shropshire) and their Brachiopoda, Pteropoda, Gasteropoda, etc. *Quarterly Journal of the Geological Society of London*, Vol. 76, 325–386.

COCKS, L R M, HOLLAND, C H, RICKARDS, R B, and STRACHAN I. 1971. A correlation of Silurian rocks in the British Isles. *Special Report of the Geological Society of London*, No. 1.

COCKS, L R M, HOLLAND, C H, and RICKARDS, R B. 1992. A revised correlation of Silurian rocks in the British Isles. *Special Report of the Geological Society of London*, No. 21, 1–32.

COLLIER, R E L, LEEDER, M R, and MAYNARD, J R. 1990. Transgressions and regressions: a model for the influence of tectonic subsidence, deposition and eustasy, with applications to Quaternary and Carboniferous examples. *Geological Magazine*, Vol. 127, 117–128.

COOK, A H, HOSPERS, J, and PARASNIS, D S. 1951. The results of a gravity survey between Clee Hills and Nuneaton. *Quarterly Journal of the Geological Society of London*, Vol. 107, 287–302.

COOPE, G R. 1975. Climatic fluctuations in north-west Europe since the last interglacial, indicated by fossil assemblages of Coleoptera. *Geological Journal Special Issue*, Vol. 6, 153–168.

COPE, K G, and JONES, A R L. 1970. The Warwickshire Thick Coal and its mining environment. *Compte Rendu 6e Congrès International Statigraphie et de Géologie du Carbonifiere*, Sheffield, 1967, 585–598.

COPE, J C W, GETTY, T A, HOWARTH, M K, MORTON, N, and TORRENS, H S. 1980. A correlation of Jurassic rocks in the British Isles. Part One: Introduction and Lower Jurassic. *Special Report of the Geological Society of London*, No. 14.

CORFIELD, S M. 1991. The Upper Palaeozoic to Mesozoic structural evolution of the North Staffordshire Coalfield and adjoining areas. Unpublished PhD thesis, University of Keele.

CORNWELL, J C. 1992. Geophysical investigations of the Worcester district. *British Geological Survey Technical Report*, WJ/92/17.

CORNWELL, J C, and ALLSOP, J. 1981. Geophysical surveys in the Atherstone district (geological mapsheet 155). *Report Applied Geophysics Unit, Institute of Geological Sciences*, No. 38.

CROSSKEY, H W. 1882. On a section of glacial drift recently exposed in Icknield Street, Birmingham. *Proceedings of the Birmingham Philosophical Society*, Vol. 3, 209.

DAVIS, A G. 1967. The mineralogy and phase equilibrium of Keuper Marl. *Quarterly Journal of Engineering Geology*. Vol. 1, 25–38.

DEPARTMENT OF THE ENVIRONMENT. 1987. Development of contaminated land, August 1987. *Circular*, 21/87. (London: HMSO.)

DEPARTMENT OF THE ENVIRONMENT. 1989. Landfill sites: Development Control. *DoE circular*, 17/89: (London: HMSO.)

DEPARTMENT OF THE ENVIRONMENT. 1990. Development on unstable land, 1990. *Planning Policy Guidance*, PPG14 (London: HMSO.)

DEPARTMENT OF THE ENVIRONMENT. 1991a. Derelict land grant policy. *Derelict Land Grant advice*, 1. (London: HMSO.)

DEPARTMENT OF THE ENVIRONMENT. 1991b. *Public registers of land which may be contaminated, a consultation paper*. (London: HMSO.)

DEPARTMENT OF THE ENVIRONMENT. 1994a. Environmental assessment. *DoE circular*, 7/94. (London: HMSO.)

DEPARTMENT OF THE ENVIRONMENT. 1994b. Treatment of disused mine openings. *Minerals Policy Guidance*, MPG12. (London: HMSO.)

DEPARTMENT OF THE ENVIRONMENT, TRANSPORT AND THE REGIONS. 1998. *Regional planning guidance for the West Midlands region. Government Office for the West Midlands*, RPG11. (London: HMSO.)

DONOVAN, D T, HORTON, A, and IVIMEY-COOK, H C. 1979. The transgression of the Lower Lias over the northern flank of the London Platform. *Journal of the Geological Society of London*, Vol. 136, 165–173.

DOWNING, R A, LAND, D H, ALLENDER, R, LOVELOCK, P E R, and BRIDGE, L R. 1970. The hydrogeology of the Trent River basin. *Water Supply Paper of the Institute of Geological Sciences, Hydrogeological Report*, No. 5.

DUIGAN, S L. 1956. Pollen analysis of the Nechells interglacial deposits, Birmingham. *Quarterly Journal of the Geological Society of London*, Vol. 112, 373–391.

DUMBLETON, M J, and WEST, G. 1966. Study of the Keuper Marl: geology and geography. *Road Research Laboratory, Ministry of Transport, Research Report*, No. 39.

DUMBLETON, M J, and WEST, G. 1966. Studies of the Keuper Marl–mineralogy. *Report of the Road Research Laboratory*, No. 40, 1–25.

DUTTON, C, GAHIR, J S, and JONES H L M. 1991. Landfill gas — experience in the Black Country. Symposium papers Methane: facing the problems, 6.1.1–6.1.4. 2nd Symposium & Exhibition. Nottingham 1991.

EASTWOOD, T, GIBSON, W, CANTRILL, T C, and WHITEHEAD, T H. 1923. The geology of the country around Coventry, including an account of the Carboniferous rocks of the Warwickshire Coalfield. *Memoir of the Geological Survey of Great Britain*, Sheet 169 (England and Wales).

EASTWOOD, T, WHITEHEAD, T H, and ROBERTSON, T. 1925. The geology of the country around Birmingham. *Memoir of the Geological Survey of Great Britain*, Sheet 168 (England and Wales).

EDMUNDS, W M, COOK, J M, KINNIBURGH, D G, MILES, D L, and TRAFFORD, J M. 1989. Trace-element occurrence in British groundwaters. *British Geological Survey Research Report*, SD/89/3.

EDWARDS, J S. 1991. Methane in groundwater. Symposium papers. Methane: facing the problems, 2.3.1–2.3.8. 2nd Symposium & Exhibition. Nottingham 1991.

EGLINGTON, M. 1979. Chemical considerations on filled ground. Proceedings of the symposium on the engineering behaviour of industrial and urban fill. *Midlands Geotechnical Society*, B11–B15.

EHLERS, J, and LINKE, G. 1989. The origin of deep buried channels of Elsterian age in NW Germany. *Journal of Quaternary Science*, Vol. 4, 255–265.

ELLIOTT, R E. 1961. The stratigraphy of the Keuper Series in southern Nottinghamshire. *Proceedings of the Yorkshire Geological Society*, Vol. 33, 197–224.

EVANS, D J, REES, J G, and HOLLOWAY, S. 1993. The Permian to Jurassic stratigraphy and structural evolution of the central Cheshire Basin. *Journal of the Geological Society of London*, Vol. 150, 857–870.

FAWDRY, J. 1913. On some stones suitable to the atmospheric conditions prevailing in Birmingham and similar districts, with some observations on their character and use.   *The Stone Trades Journal*, 215–216; 237–238; 312–314; 347–348; 374–375

FITCH, F J, MILLER, J A, and THOMPSON, D B. 1966.   The palaeogeographic significance of isotope age determinations from the Trias of the Stockport–Macclesfield district, Cheshire, England.   *Palaeogeography, Palaeoclimatology, Palaeoecology*, Vol. 2, 281–312.

FLEET, W F. 1923.   Notes on the Triassic sands near Birmingham with special reference to their heavy detrital minerals.   *Proceedings of the Geologists' Association*, Vol. 34, 114–119.

FLEET, W F. 1925.   The chief heavy detrital minerals in the rocks of the English Midlands.   *Geological Magazine*, Vol. 62, 98–128.

FLEET, W F. 1927.   The heavy minerals of the Keele, Enville, 'Permian' and Lower Triassic Rocks of the Midlands and the correlation of these strata.   *Proceedings of the Geologists' Association*, Vol. 38, 1–48.

FLEET, W F. 1929.   Petrography of the Upper Bunter Sandstone of the Midlands.   *Proceedings of the Birmingham Natural History and Philosophical Society*, Vol. 15, 213–217.

FORD, M, TELLAM, J H, and HUGHES, M. 1992.   Pollution-related acidification in the urban aquifer, Birmingham, UK.   *Journal of Hydrology*, Vol 140, 297–312.

FORSTER, A. 1991.   The engineering geology of Birmingham West (the Black Country).   *British Geological Survey Technical Report*, WN/91/15.

FOXALL, W H. 1917.   The geology of the Eastern Boundary Fault of the South Staffordshire Coalfield.   *Proceedings of the Birmingham Natural History Society*, Vol. 14, 46–54.

FRASER, A J, NASH, D F, STEELE, R P, and EBDON, C C. 1990.   A regional assessment of the intra-Carboniferous play of northern England.   417–440 *in* Classic petroleum provinces. BROOKS, J (editor). *Special Publication of the Geological Society of London*, Vol. 50.

FULTON, I M. 1987a.   Genesis of the Warwickshire Thick Coal: a group of long-residence histosols.   201–218 *in* Coal and coal-bearing strata: recent advances, SCOTT A C (editor). *Special Publication of the Geological Society of London*, No. 32.

FULTON, I M. 1987b.   The Silesian sub-system in Warwickshire, some aspects of its palynology, sedimentology and stratigraphy. Unpublished PhD thesis, University of Aston.

FULTON, I M. 1990.   Field guide: Coal Geology Group underground visit to Coventry Colliery, Warwickshire, Wednesday 7 November 1990.   Unpublished.

FULTON, I M, and WILLIAMS, H. 1988.   Palaeogeographical change and controls on Namurian and Westphalian A/B sedimentation at the southern margin of the Pennine Basin. 178–199 in *Sedimentation in a synorogenic Basin Complex: the Upper Carboniferous of northwest Europe.* BESLY, B M, and KELLING, G (editors). (Glasgow and London: Blackie.)

GALTON, P M. 1985.   The poposaurid thecodontian *Teratosaurus suevicus* v. Meyer, plus referred specimens mostly based on prosauropod dinosaurs, from the Middle Stubensandstein (Upper Triassic) of Nordwurttemburg. *Stuttgärter Beiträge zur Naturkunde*, Serie B, No. 116, 1–29.

GEIGER, M E, and HOPPING, C A. 1968.   Triassic stratigraphy of the southern North Sea Basin.   *Philosophical Transactions of the Royal Society, London*, Series B 254, 1–36.

GIBBARD, P L. 1991.   The Wolstonian Stage in East Anglia.   7–13 in *Central east Anglia and the Fen Basin: Field Guide.* LEWIS, S G, WHITEMAN, C A, and BRIDGLAND, D R (editors). (London, Quaternary Research Association.)

GIBSON, W. 1901.   On the character of the Upper Coal Measures of North Staffordshire, Denbighshire, South Staffordshire, and Nottinghamshire, and their relation to the Productive series.   *Quarterly Journal of the Geological Society of London*, Vol.58, 251–266.

GLENNIE, K W. 1990.   Rotliegend sediment distribution: a result of late Carboniferous movements.   127–138 *in* Tectonic processes responsible for Britain's oil and gas reserves. HARDMAN, R F P, and BROOKS, J (editors).   *Special Publication of the Geological Society of London*, No. 55.

GLENNIE, K W, and EVANS, G. 1976.   A reconnaissance of the Recent sediments of the Ranns of Kutch, India.   *Sedimentology*, Vol. 23, 625–647.

GLOVER, B W. 1990a.   Geology of the Halesowen district (SO98SE).   *British Geological Survey Technical Report*, WA/90/74.

GLOVER, B W. 1990b.   Geology of the Harborne district (SP08SW).   *British Geological Survey Technical Report*, WA/90/75.

GLOVER, B W. 1991a.   Geology of the Wombourne district. *British Geological Survey Technical Report*, WA/90/76.

GLOVER, B W. 1992.   Geology of the Middleton district (SP19NE). *British Geological Survey Technical Report*, WA/92/67.

GLOVER, B W, and POWELL, J H. 1996.   Interaction of climate and tectonics upon alluvial architecture: Late Carboniferous – Early Permian sequences at the southern margin of the Pennine Basin, UK.   *Palaeogeography, Palaeoclimatology and Palaeoecology*, Vol. 121, 13–34.

GLOVER, B W, POWELL, J H, and WATERS, C N. 1993.   Etruria Formation (Westphalian C) palaeoenvironments and volcanicity on the southern margins of the Pennine Basin, South Staffordshire, England.   *Journal of the Geological Society* of London, Vol. 150, 737–750.

GUION, P D, and FIELDING, C R. 1988.   Westphalian A and B sedimentation in the Pennine Basin, UK. 178–199 in *Sedimentation in a synorogenic basin complex: the Upper Carboniferous of Northwest Europe.* BESLY, B M, and KELLING, G (editors). (Glasgow: Blackie.)

GUION, P D, and FULTON, I M. 1986.   Field workshop, Daw Mill Colliery, Warwickshire,   *Field Guide, BSRG Annual Meeting, 1986, Nottingham.*

GUION, P D, FULTON, I M, and JONES, N S. 1995.   Sedimentary facies of the coal-bearing Westphalian A and B north of the Wales–Brabant High.   45–78 *in* European coal geology. WHATELEY, M K, and SPEARS, D A (editors). *Special Publication of the Geological Society of London* , No. 82.

HAINS, B A, and HORTON, A. 1969.   *British regional geology. Central England*, (3rd edition).   (London: HMSO.)

HALLSWORTH, C R. 1992a.   Stratigraphic variations in the heavy minerals and clay mineralogy of the Westphalian to ?Early Permian succession from the Daleswood Farm borehole, and the implications for sand provenance.   *British Geological Survey Technical Report*, WH/92/185R.

HALLSWORTH, C R. 1992b.   Stratigraphic variations in the heavy minerals of the Westphalian succession from the Romsley Borehole.   *British Geological Survey Technical Report*, WH/92/186R.

HALLSWORTH, C R. 1994.   Heavy mineral characterisation of Carboniferous to Triassic sandstones in the Warwickshire area and their importance for provenance.   *British Geological Survey Technical Report*, WH/94/88/R

HAMBLIN, R J O. 1982.   *Geological notes and details for 1:10 000 sheets: SO99SW (Dudley and Sedgley).*   (Keyworth, Nottingham: Institute of Geological Sciences.)

HAMBLIN, R J O. 1984. *Geological notes and details for 1:10 000 sheets: SO97SE Rubery.* Appendix A (OWENS, B.): Palynological report on two samples from the Halesowen Formation. (Keyworth, Nottingham: British Geological Survey.)

HAMBLIN, R J O, and GLOVER, B W. 1991a. Geological notes and local details for 1:10 000 sheet SO99SW (Dudley and Sedgley). (2nd edition). *British Geological Survey Technical Report,* WA/91/72.

HAMBLIN, R J O, and GLOVER, B W. 1991b. Geological notes and local details for 1:10 000 sheet SO99SE (Dudley and Wednesbury). (2nd edition). *British Geological Survey Technical Report,* WA/91/73.

HAMBLIN, R J O, and POWELL, J H. 1992. Geological notes and local details for 1:10 000 sheet SO99NW (Wolverhampton). (2nd edition). *British Geological Survey Technical Report,* WA/91/77.

HAMBLIN, R J O, HENSON, M R, and POWELL, J H. 1992. Geological notes and local details for 1:10 000 sheet SO99NE Willenhall and Darlaston. (2nd edition). *British Geological Survey Technical Report,* WA/91/78.

HAMBLIN, R J O, HENSON, M R, and WATERS, C N. 1991. Geological notes and local details for 1:10 000 sheet SP09NW (Walsall). (2nd edition). *British Geological Survey Technical Report,* WA/91/63.

HARDAKER, W H. 1912. On the discovery of a fossil-bearing horizon in the 'Permian' rocks of Hamstead Quarries near Birmingham. *Quarterly Journal of the Geological Society of London,* Vol. 68, 639-683.

HARRIS, P M, HIGHLEY, D E, HILLIER, J A, and WHITWOOD, A (compilers). 1994. *Directory of mines and quarries 1994:* (Fourth edition). (Keyworth, Nottingham: British Geological Survey.)

HARRISON, R K, OLD, R A, STYLES, M T, and YOUNG, B R. 1983. Coffinite nodules from the Mercia Mudstone Group (Triassic) of the IGS Knowle Borehole, West Midlands. *Report of the Institute of Geological Sciences,* No. 83/10, 12–16.

HASSAN, S M. 1964. A comparative study of the sedimentology of the Upper Mottled Sandstone and the Lower Keuper Sandstone of an area west of Birmingham. Unpublished MSc thesis, University of Birmingham.

HAUBOLD, H, and SARJEANT, W A S. 1973. Tetrapodenfährten aus den Keele und Enville Groups (Permokarbon: Stefan und Autun) von Shropshire und South Staffordshire, Großbritannien. *Zeitschrift für Geologische Wissenschaften,* Vol. 1, 895-933. [in German].

HEALY, P R, and HEAD, J M. 1984. Construction over abandoned mine workings. *CIRIA Special Publication,* 32, PSA. *Civil Engineering Technical Guide,* Vol. 34, 94.

HMIP (HER MAJESTY'S INSPECTORATE OF POLLUTION). 1989. The control of land-fill gas. *Waste Management Paper,* No. 27. (London: HMSO.)

HOARE, R H. 1959. Red beds in the coal measures of the West Midlands. *Transactions of the Institute of Mining Engineers,* Vol. 119, 185–198.

HOBBS, P R N. 1990. The engineering geology of the Coventry area. *In* The geology of the Coventry area. OLD, R A, BRIDGE, D McC, and REES, J G. *British Geological Survey Technical Report,* WA/89/29.

HOLDRIDGE, D A. 1959. Compositional variation in Etruria Marls. *Transactions of the British Ceramic Society,* Vol. 58, 301–328.

HOLLOWAY, S. 1985. Triassic. 31–33 in *Atlas of onshore sedimentary basins in England and Wales. Post-Carboniferous tectonics and stratigraphy.* WHITTAKER, A (editor). (Glasgow and London: Blackie for British Geological Survey.)

HORTON, A. 1974. The sequence of Pleistocene deposits proved during the construction of Birmingham motorways. *Report of the Institute of Geological Sciences,* No. 74/11.

HORTON, A. 1975. The engineering geology of the Pleistocene deposits of the Birmingham district. *Report of the Institute of Geological Sciences,* No. 75/4.

HORTON, A. 1989. Quinton. 69–77 in *The Pleistocene of the West Midlands: field guide.* KEEN, D H (editor). (Cambridge: Quaternary Research Association.)

HULL, E. 1869. The Triassic and Permian rocks of the Midland counties of England. *Memoir of the Geological Survey of Great Britain.*

HUTCHINSON, J N, SOMERVILLE, S H, and PETLEY, D J. 1973. A landslide in periglacially disturbed Etruria Marl at Bury Hill, Staffordshire. *Quarterly Journal of Engineering Geology,* Vol. 6, 377–404.

ICK, W. 1842. On some superficial deposits near Birmingham. *Proceedings of the Geological Society of London,* III, No. 89, 731–732.

ICRCL. 1988. Interdepartmental Committee on the Redevelopment of contaminated land. "Notes on the redevelopment of landfill sites". *ICRL Guidance Notes,* 17/78, May 1988.

INSTITUTE OF GEOLOGICAL SCIENCES. 1982. IGS boreholes 1980. *Report of the Institute of Geological Sciences,* No. 81/11.

IVIMEY-COOK, H C, and POWELL, J H. 1991. Late Triassic and early Jurassic biostratigraphy of the Felixkirk Borehole, North Yorkshire. *Proceedings of the Yorkshire Geological Society,* Vol. 48, 367–374.

JACKSON, D, and LLOYD, J W. 1983. Groundwater chemistry of the Birmingham Triassic sandstone aquifer and its relation to structure. *Quarterly Journal of Engineering Geology,* Vol. 16, 135–142.

JEANS, C V. 1978. The origin of Triassic clay assemblages of Europe with special reference to the Keuper Marl and Rhaetic parts of England. *Philosophical Transactions of the Royal Society of London,* A, Vol. 289, 549–639.

JOHNSON, S A. 1995. Palaeomagnetic analysis of samples from the Enville Formation, Hopwas Breccia and Kidderminster Formation of the West Midlands. *Unpublished report to BGS,* GA/94E/12.

JONES, N S. 1992. Sedimentology of the Langsettian (Westphalian A) and Duckmantian (Westphalian B) from the Coventry area of the Warwickshire Coalfield. *British Geological Survey Technical Report,* WH/92/172C.

JONES, R L, and KEEN, D H. 1993. *Pleistocene environments in the British Isles.* (London: Chapman and Hall.)

JUKES, J B. 1859. The South Staffordshire Coalfield. *Memoir of the Geological Survey of Great Britain.*

KAY, H. 1913. On the Halesowen Sandstone Series of the South Staffordshire coalfield. *Quarterly Journal of the Geological Society of London,* Vol. 69, 449.

KAY, H. 1921. From coal measures to Trias in the West Bromwich–Sandwell–Hamstead area. *Proceedings of the Birmingham Natural History and Philosophical Society,* Vol. 14, 147.

KEEN, D H (EDITOR). 1989. *The Pleistocene of the West Midlands: Field Guide.* (Cambridge: Quaternary Research Association.)

KELLING, G. 1974. Upper Carboniferous sedimentation in South Wales. 185–224 in *The Upper Palaeozoic rocks of Wales.* OWEN, T R (editor). (Cardiff: University of Wales Press.)

KELLY, M R. 1964. The Middle Pleistocene of North Birmingham. *Philosophical Transactions of the Royal Society, Series B*, Vol. 247, 533–592.

KELLY, M R, and OSBOURNE, P J. 1964. Two floras and faunas from the alluvium at Shustoke, Warwickshire. *Proceedings of the Linnaean Society, London*, Vol. 17, 37–65.

KING, W W. 1893. Clent Hills Breccia. *Midland Naturalist*, Vol. 16, 25–37.

KING, W W. 1899. The Permian Conglomerates of the Lower Severn Basin. *Quarterly Journal of the Geological Society of London*, Vol. 55, 97–127.

KING, W W. 1917. The Downtonian of south Staffordshire. *Proceedings of the Birmingham Natural History Society*, Vol. 14, 90–99.

KING, W W. 1921. The plexography of south Staffordshire in Avonian times. *Transactions of the Institute of Mining Engineering*, Vol. 61, 155–168.

KING, W W. 1923a. The Sandwell–Handsworth railway section. *Proceedings of the Birmingham Natural History and Philosophical Society*, Vol. 15, 41–46.

KING, W W. 1923b. The unconformity below the trappoid (Permian?) breccias. *Proceedings of the Worcestershire Naturalists' Club*, Vol. 8, 3–8.

KING, M J, AND BENTON, M J. 1996. Dinosaurs in the Early and Mid Triassic? — the footprint evidence from Britain. *Palaeogeography, Palaeoclimatology, Palaeocology*, Vol. 122, 213–225.

KIRTON, S R. 1984. Carboniferous volcanicity in England with special reference to the Westphalian of the East and West Midlands. *Journal of the Geological Society of London*, Vol. 141, 161–170.

KNIPE, C V, LLOYD, J W, LERNER, D N, and GRESWELL, R. 1993. Rising groundwater levels in Birmingham and the engineering implications. *Construction Industry Research and Information Association. CIRIA Special Publication*, No. 92.

LAMONT, A. 1940. Derived Upper Llandovery fossils in Bunter Pebbles from Cheadle, North Staffordshire. *Cement, Lime and Gravel*, Vol. 15, 26–30.

LAMONT, A. 1946. Fossils from Middle Bunter pebbles collected in Birmingham. *Geological Magazine*, Vol. 83, 399–44.

LAMONT, A. 1948. Illustrations of derived fossils in Middle Bunter pebbles from the Birmingham and Warwick districts. *Quarry Managers' Journal*, 3–12.

LAND, D H, 1966. Hydrogeology of the Triassic Sandstones in the Birmingham–Lichfield district. *Water Supply Papers of the Geological Survey of Great Britain, Hydrogeological Report*, No. 2.

LANDON, J. 1890. The Barr Beacon Beds. *Proceedings of the Birmingham Philosophical Society*, Vol. 7, 113–127.

LAPWORTH, C. 1913. Sketch of the geology of the Birmingham district. 10–24 in *The Birmingham country, its geology and physiography*. British Association Handbook.

LAURIE, W H. 1926. Notes on the occurrence of boulders at Selly Oak. *Proceedings of the Birmingham Natural History and Philosophical Society*, Vol. 15, 105–106.

LAWSON, J D, and WHITE, D E. 1989. The Ludlow Series in the Ludlow area. 73–90 in A global standard for the Silurian System. HOLLAND, C H, and BASSETT, M G, (editors). *National Museum of Wales Geological Series*, No. 9.

LEE, M K, PHARAOH, T C, and SOPER, N J. 1990. Structural trends in central Britain from images of gravity and aeromagnetic fields. *Journal of the Geological Society of London*, Vol. 147, 241–258.

LEE, M K, PHARAOH, T C, and GREEN, C A. 1991. Structural trends in the concealed basement of eastern England from images of regional potential field data. *Annales de la Societé Geologique de Belgique*, T. 114, 45–62.

LEEDER, M.R. 1982. Upper Palaeozoic basins of the British Isles — Caledonian inheritance verses Hercynian plate margin processes. *Journal of the Geological Society of London*, Vol. 139, 479–91.

LEEDER, M R. 1988. Recent developments in Carboniferous geology: a critical review with implications for the British Isles and N. W. Europe. *Proceedings of the Geologists' Association*, Vol. 99, 73–100.

LERNER, D N, BURSTON, M W, and BISHOP, P K, 1993. Hydrogeology of the Coventry region (UK): An urbanised, multi-layer, dual porosity aquifer system. *Journal of Hydrology*, Vol. 149, 111–135.

LESLIE, A B, TUCKER, M E, and SPIRO, B. 1992. A sedimentological and stable isotope study of travertines and associated sediments within the Upper Triassic lacustrine limestones, South Wales, UK. *Sedimentology*, Vol. 39, 613–630.

LOTT, G K. 1992a. Petrology and diagenesis of the Upper Carboniferous sandstones from the Daleswood Farm borehole, South Staffordshire. *British Geological Survey Technical Report*, WH/92/182R.

LOTT, G K. 1992b. Petrology, diagenesis and provenance of the Upper Carboniferous (Westphalian) sandstones from the South Staffordshire area. *British Geological Survey Technical Report*, WH/92/198R.

LOVELOCK, P E R, 1977. Aquifer properties of Permo-Triassic Sandstones in the United Kingdom. *Bulletin of the Geological Survey of Great Britain*, No. 56.

MADDY, D, COOPE, G R, GIBBARD, P L, GREEN, C P, and LEWIS, S G. 1994. Reappraisal of Middle Pleistocene fluvial deposits near Brandon, Warwickshire and their significance for the Wolston glacial sequence. *Journal of the Geological Society of London*, Vol. 151, 221–233.

MAROOF, S I. 1974. Geophysical investigations of the Carboniferous and pre-Carboniferous formations of the East Midlands of England. Unpublished PhD thesis, University of Leicester.

MARSHALL, C E. 1942. Field relations of certain of the basic igneous rocks associated with the Carboniferous strata of the Midland counties. *Quarterly Journal of the Geological Society of London*, Vol. 98, 1–25.

MARSHALL, C E. 1945. The Barrow Hill Intrusion, south Staffordshire. *Quarterly Journal of the Geological Society of London*, Vol. 101, 177–204.

MARSLAND, A, and POWELL, J J M. 1990. Pressuremeter tests on stiff clays and soft rocks: factors affecting measurements and their interpretation. 91–110 *in* Field testing in engineering geology. BELL, F G, CRIPPS, J C, CULSHAW, M G, and COFFEY, J R (editors). *Geological Society Engineering Geology Special Publication*, No. 6.

MARTIN, F W. 1891. The boulders of the Midland district. *Proceedings of the Birmingham Philosophical Society*, Vol. 7, 85–112.

MATLEY, C A. 1912. The Upper Keuper (or Arden) Sandstone Group and associated rocks of Warwickshire. *Quarterly Journal of the Geological Society of London*, Vol. 68, 252–280.

MAYALL, M J. 1981. The Late Triassic Blue Anchor Formation and the initial Rhaetian marine transgression in south-west Britain. *Geological Magazine*, Vol. 118, 377–384.

MCBRIDE, E F. 1963. A classification of common sandstones. *Journal of Sedimentary Petrology*, Vol. 33, 664–669.

McNestry, A. 1994. Report on the palynology of the Daleswood Farm borehole (S0 95117 79132 97 NE) of Westphalian D age. *British Geological Survey Technical Report*, WH/94/73R.

McKenzie, D P. 1978. Some remarks on the development of sedimentary basins. *Earth and Planetary Science Letters*, Vol. 40, 25–32.

Mitchell, G H. 1942. The geology of the Warwickshire coalfield in new series one-inch sheets 154–5, 168–9, 184. *Geological Survey Wartime Pamphlet*, No. 25.

Monkhouse, R A. 1986. A statistical study of specific capacities of boreholes in the Sherwood Sandstone Group around Birmingham and Wolverhampton. *British Geological Survey Technical Report*, WD/86/1.

Morgan, A V. 1973. The Pleistocene geology of the area north and west of Wolverhampton, Staffordshire, England. *Philosophical Transactions of the Royal Society, Series B*, Vol. 265, 233–297.

Morton, A C. 1992. Heavy mineral assemblages in Triassic sandstones of southern Britain and the adjacent continental shelf. *British Geological Survey Technical Report*, WH/92/206C.

Murchison, R I, and Strickland, H E. 1840. On the upper formation of the New Red Sandstone System in Gloucestershire, Worcestershire and Warwickshire. *Transactions of the Geological Society of London*, Vol. 5 (2nd Series), 331–348.

Newell Arber, E A, 1916. The structure of the South Staffs. Coalfield, with special reference to the concealed areas and to the neighbouring fields. *Transactions of the Institute of Mining Engineering*, Vol. 7, 35–70.

NACSN (North American Commission on Stratigraphic Nomenclature). 1983. North American Stratigraphical Code. *Bulletin of the American Association of Petroleum Geologists*, Vol. 67, 841–875.

Old, R A. 1982. *Geological notes and local details for 1:10 000 sheet SP 17NE (Solihull and Knowle)*. (Keyworth, Nottingham: Institute of Geological Sciences.)

Old, R A. 1983. *Geological notes and local details for 1:10 000 sheet SP 07NE (Hollywood)*. (Keyworth, Nottingham: Institute of Geological Sciences.)

Old, R A. 1987. *Geological notes and local details for 1:10 000 sheet SP 27NW (Berkswell and Balsall Common)*. (Keyworth, Nottingham: British Geological Survey.)

Old, R A. 1989. Geological notes and local details for 1:10 000 sheet: SP28NW (Maxstoke). *British Geological Survey Technical Report*, WA/89/20.

Old , R A, Sumbler, M G, and Ambrose, K. 1987. Geology of the country around Warwick. *Memoir of the British Geological Survey*, Sheet, 184 (England and Wales).

Old, R A, Bridge, D McC, and Rees, J G. 1990. Geology of the Coventry area. *British Geological Survey Technical Report*, WA/89/29.

Old, R A, Hamblin, R J O, Ambrose, K, and Warrington, G. 1991. Geology of the country around Redditch. *Memoir of the British Geological Survey*, Sheet 183 (England and Wales).

Osborne, P J, 1974. An insect assemblage of Early Flandrian age from Lea Marston, Warwickshire, and its bearing on the contemporary climate and ecology. *Quaternary Research*, Vol. 4, 471–486.

Ove Arup and Partners (Department of the Environment). 1983. *A study of limestone workings in the west Midlands*. Report to the Department of the Environment and the Metropolitan Councils of Dudley, Sandwell and Walsall, and West Midlands County Council, April 1983.

Owens, B. 1990. Palynological report on a coal sample from Little London Brook, Alveley, Shropshire. *British Geological Survey Technical Report, Stratigraphy Series*, WH/90/254R.

Parasnis, D S. 1952. A study of rock densities in the English Midlands. *Monthly Notes of the Royal Astronomical Society Journal Geophysical Supplement*, Vol. 6, 252–271.

Paton, R L. 1974. Lower Permian Pelycosaurs from the English Midlands. *Palaeontology*, Vol. 17, 541–552.

Paton, R L. 1975. A Lower Permian Temnospondylus amphibian from the English Midlands. *Palaeontology*, Vol. 18, 831–845.

Pedley, R C, 1991. *GRAVMAG — User Manual. Interactive 2.5D Gravity and Magnetic Modelling Program*. British Geological Survey, Keyworth, Nottingham.

Perrin, R M S, Rose, J, and Davies, H. 1979. The distribution, variations and origins of pre-Devensian tills in eastern England. *Philosophical Transactions of the Royal Society, London*, Series B, 287, 535–570.

Pharaoh, T C, Merriman, R J, Webb, P C, and Beckinsale, R D. 1987a. The concealed Caledonides of eastern England: preliminary results of a multidisciplinary study. *Proceedings of the Yorkshire Geological Society*, Vol. 46, 355–369.

Pharaoh, T C, Webb, P C, Thorpe, R S, and Beckinsale, R D. 1987b. Geochemical evidence for the tectonic setting of late Proterozoic volcanic suites in central England. 541–552 *in* Geochemistry and mineralization of proterozoic volcanic suites. Pharaoh, T C, Beckinsale, R D, and Rickard, D (editors). *Special Publication of the Geological Society of London*, No. 33.

Pharaoh, T C, Brewer, T S, and Webb, P C. 1993. Subduction-related magmatism of late Ordovician age in eastern England. *Geological Magazine*, Vol. 130, 647–56.

Pickering, R. 1957. The Pleistocene geology of the south Birmingham area. *Quarterly Journal of the Geological Society of London*, Vol. 113, 223–240.

Piper, D P. 1993a. Geology of the Erdington district (SP 19 SW). *British Geological Survey Technical Report*, WA/93/09.

Piper, D P. 1993b. Geology of the Stetchford district (SP 18 NW). *British Geological Survey Technical Report*, WA/93/52.

Piper, D P. 1993c. Geology of the Acock's Green district (SP 18 SW). *British Geological Survey Technical Report*, WA/93/79.

Pocock, R W. 1931. The age of the Midland basalts. *Quarterly Journal of the Geological Society of London*, Vol. 87, 1–12.

Pollard, J E. 1985. *Isopodichnus*, related arthropod trace fossils and notostracans from Triassic fluvial sediments. *Transactions of the Royal Society of Edinburgh*, Vol. 76, 273–285.

Poole, E G. 1970. Trial boreholes in coal measures at Dudley, Worcestershire. *Bulletin of the Geological Survey of Great Britain*, No. 33, 1–41.

Poole, E G. 1975. Correlation of the Upper Coal Measures of Central England and adjoining areas, and their relationship to the Stephanian of the continent. *Bulletin Societé Géologique de Belgique*, 84, 57–66

Poole, E G. 1978. The stratigraphy of the Withycombe Farm borehole, near Banbury, Oxfordshire. *Bulletin of the Geological Survey of Great Britain*, No. 68.

Powell, J H. 1984. Lithostratigraphical nomenclature of the Lias Group in the Yorkshire Basin. *Proceedings of the Yorkshire Geological Society*, Vol. 45, 51–57.

Powell, J H. 1991a. Geology of the Lye district (SO 98 SW). *British Geological Survey Technical Report*, WA/91/59.

POWELL, J H. 1991b. Geology of the Smethwick district (SP 08 NW). *British Geological Survey, Technical Report*, WA/91/71.

POWELL, J H. 1991c. Geology of the Penn district (SO 89 NE). *British Geological Survey, Technical Report*, WA/91/76.

POWELL, J H. 1993a. Geology of the Nether Whitacre district (SP 29 SW). *British Geological Survey Technical Report*, WA/92/78.

POWELL, J H. 1993b. Geology of the Birmingham City district (SP 08 NE). *British Geological Survey Technical Report*, WA/93/76.

POWELL, J H, GLOVER, B W, and WATERS, C N. 1992. A geological background for planning and development in the 'Black Country'. *British Geological Survey Technical Report*, No. WA/92/33.

PRICE, G L A. 1970. The working of limestone in the Walsall district. *Birmingham Enterprise Club Transactions*, Vol. 5, 16–35.

RAMSBOTTOM, W H C, CALVER, M A, EAGAR, R M C, HODSON, F, HOLLIDAY, D W, STUBBLEFIELD, C J, and WILSON, R B. 1978. A correlation of Silesian rocks in the British Isles. *Special Report of the Geological Society of London*, No. 10.

RAYMOND, L R. 1955. The Rhaetic beds and Tea Green Marl of North Yorkshire. *Proceedings of the Yorkshire Geological Society*, Vol. 30, 5–23.

REES, J G, and WILSON, A A. 1998. Geology of the country around Stoke on Trent. *Memoir of the British Geological Survey*, Sheet 123 (England and Wales).

RIVETT, M O, LERNER, D N, LLOYD, J W, and CLARK, L. 1990. Organic contamination of the Birmingham aquifer, UK. *Journal of Hydrology*, Vol. 113, 307–323.

ROBERTSON, T, and McCALLUM, R T. 1930. LM & SR Longbridge and Barnt Green widening near Birmingham. 42–53 in *Summary of progress for 1929. Geological Survey of Great Britain*, Pt. 2. (London: Her Majesty's Stationary Office).

ROSE, J. 1987. Status of the Wolstonian Glaciation in the British Quaternary. *Quaternary Newsletter*, Vol. 53, 1–9.

ROSE, J. 1991. Stratigraphic basis of the 'Wolstonian Glaciation' and retention of the term 'Wolstonian' as a chronostratigraphic stage name — a discussion. 15–20 in *Central East Anglia and the Fen Basin: Field guide*. LEWIS, S G, WHITEMAN, C A, and BRIDGLAND, D R (editors). (London: Quaternary Research Association.)

ROYLES, C P. 1996. Geophysical investigations in the Birmingham district. 1995. *British Geological Survey Technical Report*, WK/96/10.

R TYM and PARTNERS WITH LAND USE CONSULTANTS. 1987. *Evaluation of derelict land grant schemes*. (London: HMSO.)

RUSHTON, A W A. 1990. Note on a trilobite collected from the Enville Formation. *British Geological Survey Technical Report*, Stratigraphy Series, WA/90/234R.

RUSHTON, A W A. 1994a. Fossils from a clast in the Hopwas Breccia, Great Barr, Birmingham. *British Geological Survey Technical Report*, Stratigraphy Series, WA/94/183R.

RUSHTON, A W A. 1994b. Cambrian fossils from the Nechell Breccia. *British Geological Survey Technical Report*, Stratigraphy Series, WA/94/287R.

RUSHTON, A W A, and MOLYNEUX, S G. 1990. The Withycombe Formation (Oxfordshire subcrop) is of early Cambrian age. *Geological Magazine*, Vol. 127, 363.

RUSHTON, K R, and SALMON, S. 1993. Significance of vertical flow through low-conductivity zones in Bromsgrove Sandstone aquifer. *Journal of Hydrology*, Vol. 152, 131–152.

RUST, B R, and NANSON, C G. 1989. Bedload transport of mud as pedogenic aggregates in modern and ancient rivers. *Sedimentology*, Vol. 36, 291–306.

SARJEANT, W A S. 1996. A re-appraisal of some supposed dinosaur footprints from the Triassic of the English Midlands. *Mercian Geologist*, Vol. 14, 22–30.

SCHUMM, S A. 1977. *The fluvial system*. (New York: J Wiley and Sons.)

SCHUMM, S A. 1993. River response to base level change: implications for sequence stratigraphy. *Journal of Geology*, Vol. 101, 279–294.

SCOTESE, C R, and McKERROW, W S. 1990. Revised world maps and introduction. 1–21 in Palaeozoic palaeogeography and biogeography. McKERROW, W S, and SCOTESE, C R (editors). *Memoir of the Geological Society of London*, No.12.

SCOTT, A C. 1979. The ecology of Coal Measures flora from Northern Britain. *Proceedings of the Geologists' Association*, Vol. 90, 97–116.

SEDGWICK, A. 1829. On the geological relations and internal structure of the Magnesian Limestone, and the lower proportions of the New Red Sandstone series in their range through Nottinghamshire, Derbyshire, Yorkshire, and Durham to the southern extremity of Northumberland. *Transactions of the Geological Society of London*, Series 2, 3, 37–124.

SELLWOOD, B B, DURKIN, M K, and KENNEDY, W J. 1970. Field meeting on the Jurassic and Cretaceous rocks of Wessex. *Proceedings of the Geologists' Association*, Vol. 81, 715–732.

SHOTTON, F W. 1927. The conglomerates of the Enville Series of the Warwickshire Coalfield. *Quarterly Journal of the Geological Society of London*, Vol. 83, 604–**XXX**

SHOTTON, F W. 1929. The geology of the country around Kenilworth (Warwickshire). *Quarterly Journal of the Geological Society of London*, Vol. 85, 167–222.

SHOTTON, F W. 1953. The Pleistocene deposits of the area between Coventry, Rugby and Leamington and their bearing on the topographical development of the Midlands. *Philosophical Transactions of the Royal Society of London*. Series B, Vol. 237, 209–260.

SHOTTON, F W. 1954. The geology around Hams Hall, near Coleshill, Warwickshire. *Proceedings of the Coventry Natural History and Science Society*. Vol. 2, 237–244.

SHOTTON, F W. 1977. The English Midlands. INQUA Excursion Guide A2, 10th INQUA Congress, Birmingham.

SHOTTON, F W. 1989. The Wolston sequence and its position within the Pleistocene. 1–4 in *The Pleistocene of the West Midlands: field guide*. KEEN, D H (editor). (Cambridge: Quaternary Research Association.)

SHOTTON, F W, and OSBOURNE, P J. 1965. The fauna of the Hoxnian interglacial deposits at Nechells, Birmingham. *Philosophical Transactions of the Royal Society of London*, Series B, 248, 353–378.

SHOTTON, F W, and WEST R G. 1969. Stratigraphic table of the British Quaternary. 155–157 in Recommendations on stratigraphical usage. *Proceedings of the Geological Society of London*, No. 1656.

SMITH, A H V, and BUTTERWORTH, M V. 1967. Miospores in the coal seams of the Carboniferous of Great Britain. *Palaeontology, Special Paper*, No. 1.

SMITH, D B, BRUNSTROM, R G W, MANNING, P I, SIMPSON, S, and SHOTTON, F W. 1974. A correlation of Permian rocks in the British Isles. *Special Report of the Geological Society of London*, No. 5.

SMITH, N J P, and RUSHTON, A W A. 1993. Cambrian and Ordovician stratigraphy related to structure and seismic profiles in the western part of the English Midlands. *Geological Magazine*, Vol. 130, 665–671.

SMITHSON, F. 1931. The Triassic sandstones of Yorkshire and Durham. *Proceedings of the Geologists' Association*, Vol. 42, 125–156.

STUBBLEFIELD, C J, and BULMAN, O M B. 1927. The Shineton Shales of the Wrekin district: with notes on their development in other parts of Shropshire and Herefordshire. *Quarterly Journal of the Geological Society of London*, Vol. 83, 96–146.

STUBBLEFIELD, C J, and TROTTER, F M. 1957. Divisions of the coal measures on Geological Survey maps of England and Wales. *Bulletin of the Geological Survey of Great Britain*, No. 13, 1–5.

SUMBLER, M G. 1982. *Geological notes and local details for 1:10 000 sheet SP 28 SW (Meriden)*. (Keyworth, Nottingham: Institute of Geological Sciences.)

SUMBLER, M G, 1983. A new look at the type Wolstonian glacial deposits of Central England. *Proceedings of the Geologists' Association*, Vol. 94, 23–31.

TALBOT, M R, HOLM, K, and WILLIAMS, M A J. 1994. Sedimentation in low-gradient desert margin systems: A comparison of the Late Triassic of northwest Somerset (England) and the late Quaternary of east-central Australia. 97–117 *in* Palaeoclimate and basin evolution of playa systems: Boulder, Colorado. ROSEN, M R (editor). *Geological Society of America Special Paper*, No. 289.

TAYLOR, S R. 1983. A stable isotope study of the Mercia Mudstone (Keuper Marl) and associated sulphate horizons in the English Midlands. *Sedimentology*, Vol. 30, 11–31.

TAYLOR, K, and RUSHTON, A W A. 1971. The pre-Westphalian geology of the Warwickshire Coalfield. *Bulletin of the Geological Survey of Great Britain*, No. 35.

THORPE, R S, BECKINSALE, R D, PATCHETT, P J, PIPER, J D A, DAVIES, G R, and EVANS, J A. 1984. Crustal growth and late Precambrian–early Palaeozoic plate tectonic evolution of England and Wales. *Journal of the Geological Society of London*, Vol. 141, 521–536.

TOMLINSON, M E. 1925. River terraces of the lower valley of the Warwickshire Avon. *Quarterly Journal of the Geological Society of London*, Vol. 81, 137–169.

TOMLINSON, M E. 1935. The superficial deposits of the country north of Stratford upon Avon. *Quarterly Journal of the Geological Society of London*, Vol. 91, 423–462.

TRUEMAN, A.E. 1947. Stratigraphical problems in the Coal Measures of Great Britain. *Quarterly Journal of the Geological Society of London*, Vol. 103, lxv–civ.

TURNER, N. 1994. Westphalian D age palynomorphs from the Halesowen Formation at Cliff Quarry, near Dost Hill, Tamworth. *British Geological Survey Technical Report, Stratigraphy Series*, WH/94/234R.

VERNON, R D. 1912. On the geology and palaeontology of the Warwickshire Coalfield. *Quarterly Journal of the Geological Society of London*, Vol. 68, 587–638

WALKER, A D. 1969. The reptile fauna of the "Lower Keuper" Sandstone. *Geological Magazine*. Vol. 106, 470–476.

WALKER, T R. 1976. Diagenetic origin of continental red beds. 240–282 in *The continental Permian in Central, West and Southern Europe*. FALKE, N (editor). (Dordrecht: Holland.)

WAGNER, R H. 1983. A lower Rotliegend flora from Ayrshire. *Scottish Journal of Geology*, Vol. 19, 135–155.

WARRINGTON, G. 1970. The stratigraphy and palaeontology of the 'Keuper' Series of the central Midlands of England. *Quarterly Journal of the Geological Society of London*, Vol. 126, 183–223.

WARRINGTON, G. 1981. The indigenous micropalaeontolgy of the British Triassic shelf sea deposits. 61–70 in *Microfosils from recent and fossil shelf seas*. NEALE, J W, and BRASIER, M D (editors). (Chichester: Horwood.)

WARRINGTON, G. 1993a. Palynology report: Mercia Mudstone Group (Triassic), Jackson's brickpit, Stonebridge, near Hampton in Arden. *British Geological Survey Technical Report, Stratigraphy Series*, WH/93/189R.

WARRINGTON, G. 1993b. Palynology report: Mercia Mudstone Group (Triassic), Holly Lane brickpit, Erdington, Birmingham. *British Geological Survey Technical Report, Stratigraphy Series*, WH/93/188R.

WARRINGTON, G, AUDLEY-CHARLES, M G, ELLIOTT, R E, EVANS, W B, IVIMEY-COOK, H C, KENT, K E, ROBINSON, P L, SHOTTON, F W, and TAYLOR, F M. 1980. A correlation of Triassic rocks in the British Isles. *Special Report of the Geological Society of London*, No. 13.

WARRINGTON, G, and IVIMEY-COOK, H C. 1992. Triassic. 97–106 *in* Atlas of palaeogeography and lithofacies. COPE, J C W, INGHAM, J K, and RAWSON, P F (editors). *Memoir of the Geological Society of London*, No. 13.

WATERS, C N. 1991a. Geology of the Rowley Regis district (SO 98 NE). *British Geological Survey Technical Report*, WA/91/55.

WATERS, C N. 1991b. Geology of the Hamstead district (SP09SW). *British Geological Survey Technical Report*, WA/91/08.

WATERS, C N. 1993. Geology of the Water Orton district (SP19SE). *British Geological Survey Technical Report*, WA/92/68.

WATERS, C N, GLOVER, B W, and POWELL, J H. 1994. Structural synthesis of the S Staffordshire UK: implications for the Variscan evolution of the Pennine Basin. *Journal of the Geological Society of London*, Vol. 151, 697–713.

WATERS, C N, GLOVER, B W, and POWELL, J H. 1995. Discussion on structural synthesis of south Staffordshire, UK: implications for the Variscan evolution of the Pennine Basin, Journal, Vol. 151, 1994, 697–713. *Journal of the Geological Society of London*, Vol. 152, 197–200.

WATERS, C N, and POWELL J H, 1994. Geology of the Perry Barr district (SP09SE). *British Geological Survey Technical Report*, WA/94/16.

WESCOTT, W A. 1993. Geomorphic thresholds and complex response of fluvial systems — some implications for sequence stratigraphy. *American Association of Petroleum Geologists Bulletin*, Vol. 77, 1208–1218.

WESSEX ARCHAEOLOGY. 1996. The Thames Valley and the Warwickshire Avon. *The English Rivers Palaeolithic Project Report*, No. 1, 1994–1995. (Trust for Wessex Archaeology Ltd and English Heritage.)

WHITE, D E, and LAWSON, J D. 1989. The Přídolí Series in the Welsh Borderland and south-central Wales. 131–141 *in* A global standard for the Silurian System. Holland, C H and Bassett, M G (editors). *National Museum of Wales Geological Series*, No. 9.

WHITE, E. 1950. A fish from the Bunter near Kidderminster. *Transactions of the Worcestershire Naturalists' Club*, Vol. 10, 185–189.

WHITEHEAD, T H, and EASTWOOD T. 1927. The geology of the southern part of the South Staffordshire Coalfield. *Memoir of the Geological Survey of Great Britain*.

WHITEHEAD, T H, and POCOCK, R.W. 1947. Dudley and Bridgnorth. *Memoir of the Geological Survey of Great Britain*, Sheet 167 (England and Wales).

WHITTARD, W F. 1979. An account of the Ordovician rocks of the Shelve Inlier in west Salop and part of north Powys. *Bulletin of the British Museum (Natural History)*, Geology, No. 33, 1–69.

WILLIAMS, B J, and WHITTAKER, A. 1974. Geology of the country around Stratford-upon-Avon and Evesham. *Memoir of the Geological Survey of Great Britain*, Sheet 200 (England and Wales).

WILLS, L J. 1910. On the fossiliferous lower Keuper rocks of Worcestershire, with descriptions of some of the plants and animals discovered therein. *Proceedings of the Geologists' Association*, Vol. 21, 249–331.

WILLS, L J. 1938. The Pleistocene development of the Severn from Bridgnorth to the sea. *Quarterly Journal of the Geological Society of London*, Vol. 94, 161–242.

WILLS, L J. 1947. *A monograph of British Triassic scorpions.* Part I, 1–74, Part II, 75–137. (London: The Palaeontographical Society.)

WILLS, L J. 1956. *Concealed coalfields.* (London: Blackie.)

WILLS, L J. 1970a. The Triassic succession in the central Midlands in its regional setting. *Quarterly Journal of the Geological Society of London*, Vol. 126. 225–285.

WILLS, L J. 1970b. The Bunter Formation at the Bellington Pumping Station of the East Worcestershire Waterworks Company. *Mercian Geologist*, Vol. 3, 387–397.

WILLS, L J. 1973. A palaeogeological map of the Palaeozoic floor below the Permian and Mesozoic formations in England and Wales. *Memoir of the Geological Society of London*, No. 7.

WILLS, L J. 1976. The Triassic of Worcestershire and Warwickshire. *Report of Institute of Geological Science,* No. 76/2.

WILLS, L J. 1978. A palaeogeological map of the Lower Palaeozoic floor beneath the cover of Upper Devonian, Carboniferous and later formations. *Memoir of the Geological Society of London*, No. 8.

WILLS, L J, and SARJEANT, W A S. 1970. Fossil vertebrate and invertebrate tracks from boreholes through the Bunter Series (Triassic) of Worcestershire. *Mercian Geologist*, Vol. 3, 399–414.

WILLS, L J, and SHOTTON, F W. 1938. A quartzite breccia at the base of the Trias in a trench near Tessall Lane. *Proceedings of the Birmingham Natural History and Philosophical Society*, Vol. 16, 181–183.

WILLS, L J, WILKINS, L G, and HUBBARD, G H. 1925. The Upper Llandovery Series of Rubery. *Proceedings of the Birmingham Natural History and Philosophical Society*, Vol. 15, 67–83.

WINGFIELD, R T R. 1989. Glacial incisions indicating the Middle and Upper Pleistocene ice limits off Britain. *Terra Nova*, Vol. 1, 538–548.

WORSSAM, B C, and OLD, R A. 1988. Geology of the country around Coalville. *Memoir of the British Geological Survey*, Sheet 155 (England and Wales).

WYMER, J. 1985. The Palaeolithic sites of East Anglia. (Norwich: Geo Books.)

ZEIGLER, A M, COCKS, L R M, and MCKERROW, W S. 1968. The Llandovery transgression of the Welsh Borderland. *Palaeontology*, Vol. 11, 736–782.

FOSSIL INVENTORY

To satisfy the rules and recommendations of the international codes of botanical and zoological nomenclature, authors of cited species are listed below.

Chapter 3   Cambrian and Ordovician

Allatheca degeeri (Holm, 1893)
Burithes alatus (Cobbold, 1919)
Camenella baltica (Bengtson, 1970)
Coleoloides typicalis Walcott, 1899
Ctenopyge bisulcata (Phillips,1848)
Eurytreta sabrinae (Callaway, 1877)
Gracilitheca aequilateralis (Cobbold, 1919)
Homagnostus obesus (Belt, 1867)
Hyolithellus micans Billings, 1872
Obolus parvulus Cobbold, 1921
Orusia lenticularis (Wahlenberg, 1818)
Parabolina spinulosa (Wahlenberg, 1818)
Rhabdinopora flabelliformis socialis (Salter,1857)
Sphaerophthalmus humilis (Phillips,1848)
Torellella biconvexa Missarzhevsky, 1969
Torellella lentiformis (Syssoiev, 1962)
Tuojdachithes biconvexus (Cobbold, 1919)

Chapter 4   Silurian

Bumastus barriensis Murchison, 1839
Costistricklandia lirata (J de C Sowerby, 1839*)*
Costistricklandia lirata 'alpha' St Joseph, 1935
Costistricklandia lirata typica = C. lirata lirata   (J de C Sowerby, 1839)
Stricklandia laevis (J de C Sowerby, 1839)

Stricklandia lens (J de C Sowerby, 1839)

Chapter 5   Carboniferous and Lower Permian

Avicula quadrata J de C Sowerby, 1840
Cadiospora magna Kosanke, 1950
Conularia quadrisulcata J Sowerby, 1821
*Cordaites brandlingi (*Lindley & Hutton) Goeppert, 1888
Dadoxylon kayi Arber, 1913
Dasyceps bucklandi (Lloyd) Huxley, 1859
Haptodus grandis Paton, 1974
Lingula mytiloides J Sowerby, 1813
Mooreisporites inusitatus (Kosanke) Neves, 1958
Orbicula nitida
Productus scabricula (Fischer de Waldheim, 1837)
Sphenacodon brittanicus (von Huene) Paton, 1974
Torispora securis (Balme) Alpern, Doubinger & Horst, 1837)

Chapter 6   Triassic

Alisporites parvus de Jersey, 1962
Alisporites toralis (Leschik) Clarke, 1965
Brodispora striata Clarke,1965
Camerosporites secatus Leschik emend. Scheuring, 1978
Chlamys valoniensis (Defrance, 1825)
Cuneatisporites radialis Leschik emend. Scheuring, 1978
Duplicisporites granulatus Leschik emend. Scheuring, 1970
Duplicisporites verrucosus Leschik emend. Scheuring, 1978
Ellipsovelatisporites plicatus Klaus, 1960
Enzonalasporites vigens Leschik, 1955
Eotrapezium concentricum (Moore,1861)
Eotrapezium germari (Dunker,1846)
Euestheria minuta (Zieten, 1833)
Gyrolepis alberti Agassiz, 1835

Haberkornia gudati Scheuring, 1978
Klausipollenites devolvens (Leschik) Clarke, 1965
Labiisporites granulatus Leschik, 1956
Lyriomyophoria postera (Quenstedt, 1856)
Ovalipollis pseudoalatus (Thiergart) Schuurman,1976
Parvisaccites triassicus Scheuring,1978
Patinasporites densus Leschik emend. Scheuring, 1970
'Natica' oppelii Moore, 1861
Permichnium völckeri Güthorl, 1934
Plaesiodictyon mosellanum Wille, 1970
Porcellispora longdonensis (Clarke) Scheuring emend. Morbey, 1975
Protocardia rhaetica (Merian, 1853)
Pteromya tatei (Richardson and Tutcher, 1914)
Rhaetavicula contorta (Portlock, 1843)
Rhaetogonyaulax rhaetica (Sarjeant) Loeblich and Loeblich emend. Below 1987
Ricciisporites umbonatus Felix and Burbridge, 1977
Rimaesporites potoniei Leschik, 1955
Spiritisporites spirabilis Scheuring, 1970
Stricklandia lens (J de C Sowerby, 1839)
Triadispora plicata Klaus emend. Scheuring, 1978
Vallasporites ignacii Leschik, 1955
Yuccites vogesianus Schimper and Mougeot, 1844

Chapter 7   Jurassic

'Ammonites planorbis' J de C Sowerby, 1824
Cardinia ovalis (Stutchbury, 1842)
Eodiadema minutum (J Buckman, 1844)
Modiolus minimus J Sowerby, 1818
Pseudolimea hettangiensis (Terquem, 1855)
Psiloceras planorbis (J de C Sowerby, 1824)
Pteromya tatei (Richardson and Tutcher, 1914)

INDEX

BRITISH GEOLOGICAL SURVEY

Keyworth, Nottingham NG12 5GG
0115 936 3100

Murchison House, West Mains Road, Edinburgh EH9 3LA
0131 667 1000

London Information Office, Natural History Museum
Earth Galleries, Exhibition Road, London SW7 2DE
020 7589 4090

The full range of Survey publications is available through the
Sales Desks at Keyworth and at Murchison House, Edinburgh,
and in the BGS London Information Office in the Natural
History Museum (Earth Galleries). The adjacent bookshop
stocks the more popular books for sale over the counter. Most
BGS books and reports can be bought from The Stationery
Office and through Stationery Office agents and retailers.
Maps are listed in the BGS Map Catalogue, and can be bought
together with books and reports through BGS-approved
stockists and agents as well as direct from BGS.

*The British Geological Survey carries out the geological survey of Great
Britain and Northern Ireland (the latter as an agency service for the
government of Northern Ireland), and of the surrounding continental
shelf, as well as its basic research projects. It also undertakes
programmes of British technical aid in geology in developing countries
as arranged by the Department for International Development and
other agencies.*

*The British Geological Survey is a component body of the Natural
Environment Research Council.*

Published by The Stationery Office and available from:

The Publications Centre
(mail, telephone and fax orders only)
PO Box 276, London SW8 5DT
Telephone orders/General enquiries 0870 600 5522
Fax orders 0870 600 5533

www.tso-online.co.uk

The Stationery Office Bookshops
123 Kingsway, London WC2B 6PQ
020 7242 6393 Fax 020 7242 6412
68–69 Bull Street, Birmingham B4 6AD
0121 236 9696 Fax 0121 236 9699
33 Wine Street, Bristol BS1 2BQ
0117 926 4306 Fax 0117 929 4515
9–21 Princess Street, Manchester M60 8AS
0161 834 7201 Fax 0161 833 0634
16 Arthur Street, Belfast BT1 4GD
028 9023 8451 Fax 028 9023 5401
The Stationery Office Oriel Bookshop
18–19 High Street, Cardiff CF1 2BZ
029 2039 5548 Fax 029 2038 4347
71 Lothian Road, Edinburgh EH3 9AZ
0870 606 5566 Fax 0870 606 5588

The Stationery Office's Accredited Agents
(see Yellow Pages)

and through good booksellers